Benchmarking in Food and Farming

Benchmarking in Food and Farming

Creating Sustainable Change

LISA JACK

Routledge
Taylor & Francis Group

LONDON AND NEW YORK

First published in paperback 2024

First published 2009 by Gower Publishing

Published 2016 by Routledge
4 Park Square, Milton Park, Abingdon, Oxon OX14 4RN

and by Routledge
605 Third Avenue, New York, NY 10158

Routledge is an imprint of the Taylor & Francis Group, an informa business

Publisher's Note
The publisher has gone to great lengths to ensure the quality of this reprint but points out that some imperfections in the original copies may be apparent.

British Library Cataloguing in Publication Data
Benchmarking in food and farming : creating sustainable
  change. -- (Gower sustainable food chains series)
  1. Agricultural productivity--Evaluation. 2. Sustainable
agriculture--Economic aspects.
  I. Series II. Jack, Lisa.
  338.1'6-dc22

Library of Congress Cataloging-in-Publication Data
Benchmarking in food and farming : creating sustainable change / [edited] by Lisa Jack.
    p. cm. -- (Gower sustainable food chains series)
  Includes index.
  ISBN 978-0-566-08835-3 (hbk)
  1. Agricultural industries--Management. 2. Food industry and trade--
Management. 3. Benchmarking (Management) I. Jack, Lisa.
  HD9000.5.B437 2009
  630.68'5--dc22
                                                                    2009016186

ISBN: 978-0-566-08835-3 (hbk)
ISBN: 978-1-03-283810-6 (pbk)
ISBN: 978-1-315-56904-8 (ebk)

DOI: 10.4324/9781315569048

Contents

List of Figures

List of Figures

List of Tables

Acknowledgements

I would like to acknowledge the help and encouragement of a number of people in the writing this book. Firstly the contributors, who kindly wrote the case-study chapters based on their greater experience of these areas than mine. I would also like to thank the many farmers I have interviewed who wished to remain anonymous and the many professionals who have taken part in the projects that I have conducted since 2001. In particular, I would like to thank Jamie Gwatkin of the Joint Venture Farming Group and Louise Letnes, Kent Olsen and Dale Nordquist of the University of Minnesota for their time and input. Acknowledgements are also due to the Food Chain Centre at the IGD for use of their outputs and for many discussions. The publishers and the editor of the series, Andrew Fearne, have also been instrumental in getting the book off the ground. Finally, my family are due much gratitude for their patience and encouragement during the writing of this book.

List of Contributors

Lisa Jack is Senior Lecturer in Accounting in the Essex Business School at the University of Essex, England. She is a qualified accountant who moved into academic life following 10 years as an auditor. Her key areas of research are accounting in the agri-food industry and performance measurement, which she began looking at whilst working at Writtle College, a partner institution of the University of Essex.

Julie Boone is a principal lecturer at Harper Adams University College in Shropshire, England. Her areas of special interest include marketing and supply chain management in the food industry. She has worked with the Food Chain Centre in the UK on its Balanced Scorecard Project and in facilitating learning groups.

Sue Kilpatrick is an associate professor and Director of the Department of Rural Health at the University of Tasmania. Her research interests are rural health systems, community participation in health, social capital, rural leadership and vocational education and training in regional Australia. She has published extensively in these areas, besides working as a consultant and researcher with rural communities at the local level.

Nicola Shadbolt is an associate professor in the agribusiness section of the Institute of Food Nutrition and Human Health at Massey University in New Zealand. She has researched and written extensively on the need for strategic management practices in agriculture and food, including benchmarking and balanced scorecards. She is the editor and an extensive contributor (with Sandra Martin) to *Farm Management in New Zealand* published by Oxford University Press in 2005.

Sarah Thelwall is a business development consultant to creative businesses and one of the founders of www.mycake.org. She has also worked as a consultant with David Thelwall of Prospect Management Services on the use and promotion of benchmarking in agriculture.

Peter Whitehead joined IGD in 2001 following 20 years' experience working at the Ministry of Agriculture, Fisheries and Food (now Defra). He has a wealth of experience on agriculture matters of policy, including the department's role in competition policy. At IGD, Peter managed the Food Chain Centre from its inception and has co-authored a number of publications, including *Lean Operations* and *Competing in the 21st Century*.

1 Sustainable Change and Benchmarking in the Food Supply Chain

LISA JACK and JULIE BOONE

Introduction

Those outside agriculture are usually very surprised to find that benchmarking is a long-standing and highly developed practice in the agricultural industry. Certainly, in reading about the development of benchmark setting and benchmarking for best practice, there is very rarely any mention of farming and food. Yet benchmarking practices can be traced back to the late nineteenth century and new, innovative practices are being developed today. A review of benchmarking in agriculture and a discussion of its future potential is overdue, particularly at a time when food producers have to make significant changes to their business practices in order to survive.

What is equally surprising is the evidence that relatively very little benchmarking takes place downstream from producers in the food supply chain, among the processors, manufacturers, distributors and retailers of the food and drinks industry.

The discussions and case studies in this book should be of interest to anyone involved in benchmarking as a practice, as well as to those involved in the food and farming industry. These are stories of innovation and collaboration among producers and their advisors. What sets benchmarking in agriculture above the usual explorations of benchmarking for best practice in major corporations is that benchmarking here is among large numbers of small, often family owned, businesses working in a global industry. This is about accounting practice as sharing, trust and a sociable activity, driven by a thirst for information. It is also about sustaining farming and food in individual businesses and for everyone: as the slogan (attributed to American farmer and philosopher Wendell Berry) has it 'if you eat, you're in agriculture'.

What is Sustainable Change?

The term 'sustainable change' is used here in its rather literal senses of making changes that can be sustained and of making changes that will sustain the business or individual who has made the decision to change. 'Sustain' has a number of connotations related to endurance, nourishment and capability of survival, all integral to the mindset of farmers and producers. On wider, societal and global levels, there are connotations of environmental protection, preservation of landscape and economic well-being. Agricultural producers as much as, and probably more than, any other group are under pressure to perform economically by creating livelihoods and profits for themselves, and affordable food, fibre and fuel for others, whilst at the same time minimizing damage to the environment. Furthermore, direct links between producers and consumers exist but are not the main means of food distribution. Producers upstream are linked through processors, distributors and retailers to consumers in sometimes complex supply chains, which themselves need to be sustainable.

What is Benchmarking?

In fourth century BC Greece, Xenephon wrote:

> *You (Socrates) have discovered the reasons why some farmers are so successful that husbandry yields them all they need in abundance, and others are so inefficient that they find farming unprofitable. I should like to hear the reasons in each case, in order that we may do what is good and avoid what is harmful.*

The term 'benchmarking' is used to cover a number of practices found in farming and food that are designed to highlight the good and make it possible to avoid the harmful. The first task is to clarify what is meant by benchmarking in each of these practices, before assessing their contribution to the development of sustainable change in these industries.

In general business practice, benchmarking is used to signify a particular systematic approach in which a business evaluates its own operations and procedures through a detailed comparison with those of another business, in order to establish best practice and to improve performance. This is usually referred to as 'best practice benchmarking'. The leading company for this fashion is often cited as being Xerox Corporation in the late 1970s. However, as shown in Part 2 of this book, agriculture has a claim to have been using a similar approach about two decades before this, through farm discussion groups. Furthermore, these groups (also referred to as business improvement clubs) involve a high level of collaboration and networking, and a notion of benchmarking as a learning practice, which in other sectors is regarded as recent and highly innovative.

Benchmarking for best practice has two forms, identified by Spendolini (1992): a *functional or generic* form, where the benchmarking activity involves the identification of products, services and work processes of external organizations that may not be in direct competition with the organization in question, involving some collaborative sharing of data; and a *competitive* form, where the products, services and processes are from direct competitors. Benchmarking activities may also be internal, where the processes and services are compared across divisions or departments or some other sub-unit, in order to set standards or to promote internal competition. In farming and some food companies, benchmarking for best practice takes a generic form – very few, if any, examples have been identified of competitive benchmarking from available data – but there are examples of internal benchmarking in corporate farm groups.

Bogan and English (1994) distinguish carefully between benchmarking for best practice and the practice of setting benchmarks for the analysis of the performance of an entity. This distinction is essential in understanding the use of benchmarks in the agricultural sector. In its widest sense, the migration of the terms 'benchmark' and 'yardstick' from surveying into business since the 1920s signifies the provision of a reference point or standard against which the performance of a business can be judged. In developed countries, the use of pooled data through surveys or client bases to provide standardized data against which financial and production performance can be measured is widespread and highly institutionalized in the agricultural sector, and has been since the late nineteenth century. The data and the performance measures derived from the data are used at local levels in farm advisory work and at national and international levels to inform agricultural policy. Academics have used the term 'comparative analysis' to distinguish the use of benchmarks for diagnostic purposes from the practice of benchmarking for best practice, which involves the analysis of processes in an entity.

In this book, benchmarking practices will be identified under three headings: *benchmarks for performance analysis; benchmarking for better practice;* and *best practice benchmarking or process benchmarking.* Underlying the whole, however, is the notion that benchmarking covers any activity where managers compare their practices and performance with someone else's, and make changes intended to result in improvement of both. This can result from something as straightforward as a walk around someone else's farm (and farm visits are considered here as an important element of benchmarking) or as complex as a spreadsheet to analyse detailed financial and physical data supplied by a number of producers.

The Link between Sustainable Change and Benchmarking

Bogan and English (1994: 211) in their book *Benchmarking for Best Practices: Winning Through Innovative Adaption* claim, 'If organizations decide to approach the management of change in a systematic way, benchmarking is arguably the single most powerful tool within their grasp.' They state that benchmarking can:

- create motivation for change;
- provide a vision for what an organization can look like after change;
- provide data, evidence, and success stories for inspiring change;
- identify best practices for how to manage change;
- create a baseline or yardstick by which to evaluate the impact of earlier changes.

Through these means, the innovations of others can be adapted to the circumstances of other businesses, and new innovations uncovered. The points listed here have been extracted from a longer list, but the other pointers they give are more relevant to larger organizations motivating a large number of employees in the change process – less of an issue in producer businesses, which are essentially small businesses.

Hence, numeric benchmarks can provide a 'kick' to producers to convince them that change is needed. The many research findings and anecdotes available show – more effectively than numbers – that seeing how other outfits work and how they have achieved successful change in their operations are great motivators for making sustainable changes.

Throughout the book, the various approaches to benchmarking are assessed for how well they meet the above criteria, and, furthermore, the extent to which the different approaches drive new and adaptive innovation.

Benchmarking in the Food Supply Chain

Recent research into the use of benchmarking in the food supply chain among distributors, manufacturers and retailers (that is, not the primary producers) has shown that there is relatively little evidence of benchmarking, despite its popularity in other commercial sectors. A CIMA sponsored study by Luther and Abdel-Kader (2006: 344) stated that 'neither EVA® nor benchmarking have yet gained popularity despite the Benchmarking and Self-Assessment Initiative', the latter being launched by Leatherhead Food Research Association in the 1990s. Furthermore, they conclude by supporting the findings of an earlier study into the area (by Mann et al., 1999) that 'UK food sector companies had less well developed management systems than other industries and were not as good at meeting financial targets using appropriate financial and non-financial indicators' (Luther and Abdel-Kader, 2006: 350).

Food Manufacture magazine carried an article reporting on a study into value adding practices in the cereals supply chain:

> *Walking the value chain from plough to plate is a great way of highlighting non value-added activity, and there is a lot of it in this sector. The complexity of the cereals supply chain is incredible, which can lead to delivery errors, transport inefficiencies, no shows, rejections on the basis of quality problems and overstocking … The information flow is consistently inefficient, the workforce is resistant to change and there are low levels of staff empowerment, recognition and reward, plus a general lack of benchmarking and electronic commerce.*

> (Barnes, 2007)

Another study from 2006, again looking at cereals, highlighted the inherent problems in supply chains as a deeper problem than the techniques available to suppliers and customers:

> *Supply chains have a good size structure providing a stable supply of raw material, well-located for usage outlets. However there appears to be a lack of co-operation, integration and trust within the supply chains. Specific training and education is required and there is a problem of recruiting and retaining skilled staff. There is a lack of benchmarking and sharing of good practice.*

> (Thewell and Ritson, 2006)

There are examples of what could loosely be termed benchmarks being used in the industry in the form of key performance indicators and standards to which suppliers are required to conform. A number of these are used as illustrations in Chapter 7 (Standards, Indices and Targets for Environmental Performance) and Chapter 9 (Process Benchmarking and other Developments in Agriculture). The sources of these standards and indicators derive from the areas of logistics (the SCOR initiative), sustainability (the supermarkets Wal-Mart and Tesco both introduced environmental standards for suppliers and themselves in 2008, for example), efficient customer response (ECR), value chain analysis and product quality. However, these are not referred to as benchmarking as such and do not conform to the definition of best practice benchmarking. Having said that, there are a number of large corporations that do employ benchmarking in parts of their business and well-known examples of this, which are illustrated in Chapter 9, include Frito-lay, Mars and Kellogg's.

Therefore, the question could be asked whether benchmarking for sustainable change is relevant throughout the food supply chain. In Chapter 9 it is argued that the data collected and available in the chain already presents opportunities for benchmarking activities that might create value to all parties, but that issues of trust and transparency need to be addressed before this potential can be realized.

Benchmarking in Farming

Despite the low take-up of benchmarking in the food supply chain downstream from primary production, there is a wealth of evidence of the use and the potential of benchmarking among farmers and their advisors. The aim of this book is straightforward: to review and evaluate the widespread existing practices in benchmarking in the industry, and then to explore the potential for further development and innovation through isolated examples of innovative practice in agriculture and other industries.

The history of benchmarking in agriculture shows that it was introduced by agricultural economists and has been extended by academics, farm management consultants and advisors, agricultural extension workers and farmers themselves over the course of a century. A significant part of benchmarking in the industry has been based on the comparative analysis of financial accounting records of groups of farmers, complemented by physical stock and husbandry records. Benchmarking has had a very numeric feel to it in the industry, but there is also evidence that more tangible, process analysis elements are being incorporated into practices. However, as will be seen in Chapter 2, the identification of best practices from records and observation can be traced back to the Middle Ages in Europe and beyond, through model farms and demonstrations of farming methods and machinery to the present day: seeking for best practice, as the quote above from Xenephon suggests, is embedded in agricultural practice.

The material in this book is drawn from a number of sources. These include historical records, academic research, the agricultural press, secondary data, the author's own original interview and observation research data, and case studies from those involved in facilitating, promoting and studying benchmarking and performance measurement in farming and food. However, it is not a comprehensive and definitive review. Leading players in the field have been identified, but not every scheme in every country is covered. This is in part owing to the research focus of the author, which has mainly encompassed the UK, the US, Australia and New Zealand. This is not to say that benchmarking does not take place elsewhere; only that the scope of this book is limited to these countries.

The studies and examples chosen are indicative of what is happening in the industry, and the aim is to create a picture of where the industry is at this time, with examples of leading-edge practice to demonstrate what is possible in the industry. The use of benchmarking among primary producers – farmers – shows clearly that there is an appetite for learning through information and that the benefit of practices already in place is the potential for learning and improvement that benchmarking brings. This aspect of benchmarking is well studied outside the industry and this book aims to highlight how farmers are becoming 'information rich' as one New Zealand study puts it (Verissimo and Woodford, 2005).

Benchmarking and Learning

Learning is a process which can involve experience, practice or insight. These bring about relatively permanent changes in ways of thinking, attitudes, knowledge, skills or behaviour. Vicarious learning can occur through observing the actions of others and their consequences, providing the knowledge can be transferred to a different situation. So benchmarking is essentially a vicarious learning tool, where imitating successful firms, or avoiding the actions of unsuccessful ones, can enhance the capacity for and speed of learning.

Benchmarking's purpose is, firstly, to help firms identify what they need to change to improve their performance. Secondly, it should provide a model or principles to guide the implementation of practices and bridge the gap between goals and aspirations. Just as people and firms learn, then so too should supply chains. It is difficult enough for people to learn; how much more of a challenge, then, for a complex network of individuals and their organizations?

As the pace of business change accelerates, speed of learning is increasingly an issue. Some companies are measuring 'capacity for learning' (Senge, 1990) – the combination of ability and attitude that allows an organization 'to learn and adapt faster than its rivals'. As benchmarking in the industry develops, a movement from mass participation comparative analysis, which is a comparatively time consuming process, towards process-based benchmarking between a small number of partners should be seen in the industry as the capacity for learning is enhanced and refined.

Benchmarking that concentrates on 'emphasizing the learning element in benchmarking, and using benchmarking as a lever for learning, can support the development of a culture of continuous improvement' according to the UK's Public Sector Benchmarking Service. In fact, there are many definitions of benchmarking, but most include learning, information sharing, and the adoption of best practices to improve performance.

Benchmarking in Farming and Food

The most established practice in farming and food is the use of mass participation datasets based on financial accounting data to produce benchmark figures which are used by advisors, policy makers and farmers themselves as an aid to problem diagnostics and business planning. The story of mass participation benchmarking is covered in Chapter 2 and questions are raised about whether this embedded practice can be said to aid sustainable change in the industry, and about its potential future use. The wealth of data available and the uses to which it is put are largely unrecognized outside the industry. In Chapter 3, the findings of a study into best practice in mass participation benchmarking are discussed, looking at a small number of highly regarded systems out of the many hundreds available worldwide. A detailed case study of the development of a new benchmark system in New Zealand – DairyBase – that is based on the experience and knowledge of best practice across the industry forms the basis of Chapter 4.

The second area relating to current practice that is examined is the development of small group benchmarking in agriculture, dating from the late 1940s in New Zealand and diffusing across Australia, the UK and very recently North America. Sometimes known as business improvement clubs or benchmarking clubs, the basis on which they operate is a small-scale version of the mass participation benchmark data covered in Part 1, with farmers sharing financial data through a facilitator. However, they go further because through discussion, site visits and research, the members of the group learn from each other's experience to make tangible changes in their operations. These groups present powerful learning opportunities and go further than mass participation benchmark results because individual processes and operations can be analysed. Using the rationale from Bogan and English (1994) above, they provide 'a vision for what an organization can look like after change' and 'data, evidence, and success stories for inspiring change'. The history and examples of small group benchmarking in agriculture are covered in Chapter 5, whilst Chapter 6 reports on the first five years of the UK project led by the Food Chain Centre to encourage the growth of business improvement clubs in agriculture and horticulture.

The most problematic area is that of environmental benchmarking in the industry. The word 'sustainable' is becoming synonymous with environmental protection and conservation. As landowners and users, primary producers are seen as being inherently responsible for land quality, which is undeniably necessary for the long-term viability of land-based food production. In Chapter 7, some of the many examples of key performance indicators, targets and standards that are being applied in agriculture and elsewhere in the food chain are investigated, but the question is asked about how best practice benchmarking can be implemented in the area of environmental preservation. Using business improvement groups is one approach to the problem, and the experiences of such groups in Australia, where the problems investigated were environmental rather than purely financial or production-based, are discussed in Chapter 8.

In the final part of the book, a number of cases of the use of more competitive forms of benchmarking taking place on an individual business level in agriculture and food are explored. These include examples of internal benchmarking in corporate farm and food companies, process benchmarking in machinery/labour rings and veterinarian issues, and the use of non-financial performance indicators in benchmarking. The potential for developing more sophisticated process benchmarking activities alongside the existing mass participation and small group approaches is discussed in Chapter 9, drawing on findings about what makes successful benchmarking practice in other sectors. The development of process benchmarking in agriculture in particular has the potential for creating sustainable change that drives both new innovation and innovative adaption of the ideas of others. One potential innovation in the practice of strategic management in farming and food is the use of the balanced scorecard and a detailed discussion about its development, based on the work of Nicola Shadbolt in New Zealand, is found in Chapter 10.

Conclusion

The use of benchmarks and the development of new forms of benchmarking for best practice is well established in agriculture, although less so in other areas of the food chain. In writing this book, the aim is both to be informative about the many practices happening and to identify the potential for future innovations in the practice of benchmarking that draw on the many strengths (and avoid some of the pitfalls) of current practices in the industry.

References

Barnes, C. (2007), Study exposes weak links in grain chain, *Food Manufacture* 27 March, 2007. <http://www.foodmanufacture.co.uk/news/fullstory.php/aid/4483/Study_exposes_weak_links_in_grain_chain.html>.

Bogan, C.E. and English, M.J. (1994), *Benchmarking for Best Practices: Winning Through Innovative Adaption* (New York: McGraw-Hill).

Luther, R. and Abdel-Kader, M. (2006), Management accounting practices in the British food and drinks industry, *British Food Journal* Vol. 108 No. 5, 336–57.

Mann, R., Adebanjo, O. and Kehoe, D. (1999), Best practices in the food and drink industry, *Benchmarking for Quality Management and Technology* Vol. 5 No. 3, 184–99.

Spendolini, M.J. (1992), *The Benchmarking Book* (New York: American Management Association).

Senge, P. (1990), *The Fifth Discipline: The Art and Practice of the Learning Organisation* (London: Random House).

Thelwell, D. and Ritson, C. (2006), The international competitiveness of the UK cereals sector. Paper prepared for presentation at the 98th EAAE Seminar 'Marketing Dynamics within the Global Trading System: New Perspectives', Chania, Crete, 29 June–2 July, 2006.

Veríssimo, A. and Woodford, K. (2005), Top performing farmers are information rich: case studies of sheep and cattle farmers in the South Island of New Zealand. Published in the Proceedings of the Fifteenth International Farm Management Association Congress, Campinas, Brazil, August 2005, Vol. 1, 365–8.

Benchmarks for Comparing Farm Performance

2 Benchmarks and Yardsticks: Mass Participation Benchmarking in Farming

Introduction

In Chapter 1, benchmarking is described in its widest sense as any activity where managers compare their practices and performance with someone else's, and make changes intended to result in the improvement of both. The setting of standards, targets and benchmarks in agriculture goes back a very long way, beginning in the Middle Ages in Europe. The aim in setting benchmarks for comparative analysis of the performance of different farms is largely diagnostic: to identify problems and areas of weak performance and to examine how high-performing farms achieve their results, with the aim of improving practice on less well performing farms. This analysis was most frequently carried out by overseers of large estates in the early history, and then by advisors and consultants providing a service to farm businesses. In order for there to be a representative set of figures, a significant number of farms are needed to pool their results to provide the comparisons. For this reason, the term 'mass participation benchmarking' has been coined to describe these activities.

The results of mass participation benchmarking are used by farmers (often through their accountants, advisors and consultants), but are also used for research purposes by agricultural economists and farm management specialists and to inform policy makers. Benchmark data sets are maintained on a local level by private firms (Grant Thornton, the accounting firm, maintains a client database, for example, in its agricultural practice and regularly publishes its data in an annual Farm Income Survey), on a wider level by state-sponsored schemes (for example, based at the land grant universities in the US or state governments in Australia), on a national level, such as the Farm Business Survey in the UK or the Australian Bureau of Agricultural and Resource Economics) and internationally (notably the Farm Advisory Data Network).

The problem of assessing the extent to which the results of mass participation benchmarking drive change through the comparative analyses obtained is simply that the key defining feature of the data sets is that the data is supplied anonymously. Results can be filtered to give individual participants a report of their performance against peer group performance but these reports are confidential. There is almost no empirical data available to make an assessment of how the information is used by individual farmers in making decisions, nor about the long-term success of changes made on the basis of those decisions, due to the anonymity. The longevity of the practice suggests that the information is used, but the extent to which the data is reliable and useful has been much debated, as will be seen in the discussion which follows.

The aim of this chapter is to provide an overview of the historical background to the practice of mass participation benchmark setting and to evaluate its role in supporting sustainable change in farming businesses. Chapter 3 is a study into 'best practice' in mass participation benchmarking, reviewing a number of schemes in different countries that collect and provide information as a service to farmers, advisors and policy makers. Chapter 4 also covers best practice in this type of benchmarking and addresses the problem of acquiring high-quality, reliable data from producers by using a standard form of input by accountancy firms into the system and having detailed, standardized outputs for dairy producers. The scheme was carefully devised using the experiences and knowledge of many players in the industry to create a state-of-the-art system.

The Early Use of Standards in the Analysis of Farm Performance

The systematic diagnosis of problems in farming using accounting has a surprisingly long history. Early records of estate management in the Middle Ages show that stewards, auditors and bailiffs working for the medieval manors in Britain had highly developed methods of measuring production capacity, setting production standards and targets to be met, operating standard cost and cost allocation systems and carrying out analyses of performance, as well as making decisions using relevant cost concepts. This was well before the usually accepted starting point for such practices in manufacturing during the Industrial Revolution. Scorgie (1997) identifies a number of records which show that good practice entailed making comparisons of performance year on year and seeking explanations for variances from standards, in order to rectify performance. One example given is from advice to the Countess of Lincoln on estate management from one Robert Grosseteste, in AD 1240:

> At the end of the year, when all the accounts of the manors have been rendered and heard, concerning all the lands, yields and expenses, collect all the rolls. Then in great secrecy, by one or two of the most discreet and faithful men you have, compare the account with the rolls containing the estimates of corn and stock which you had compiled after the previous harvest. According to how far they agree you will see the good intentions or shortcomings of your servants and bailiffs and you will be able to put things right, where necessary.

(Ibid. 52)

Following the collapse of the feudal system, when yeomen and farmers had control of their own farms or were tenants rather than serfs of the manor, such practices were left to the stewards of the great estates. Despite numerous farm accounting treatises in the interim, there is little evidence of performance analysis being carried out on farms, other than perhaps model farms on the large estates. Model accounts were suggested by Young in 1797 (Scorgie, 1997), but it was not until the end of the nineteenth century that the groundwork for the wide-scale provision of standardized data and benchmarks was established. This was first through the large-scale collection of farm accounting data to ascertain standards of performance, and secondly through the development of financial ratios to analyse the data provided by the large-scale databases and to provide benchmarks of performance.

Surveys, Statistics and Standards using Farm Accounts

Government economists began to collect physical data about farms through agricultural census in the mid-nineteenth century in the US, the UK, Germany and elsewhere. By the end of the nineteenth century, however, information about the profitability of farms and questions about whether the management of the farm was a key indicator of success began to drive the promotion of farm accounting and the large-scale collection of farm accounts to form databases for statistical analysis in a number of countries. Three different approaches to this task were taken:

- to teach farmers to keep financial accounts in a particular format and to collate the accounts of all farms at a central office. A variation on this was to have a full-time employee of the office or farmers' collective visit each farm to prepare the accounts;
- to carry out surveys, where the economist or farm management advisor visited the farm to collect the accounting data required in a standard format;
- to prepare cost accounts for farms which are then kept by the researcher.

Data Collection through Accounting Services

The first approach was exemplified by the work of one Professor Ernst Laur in Switzerland. According to Hinrichs (1929), in 1898 Laur, in his role as an economic advisor with the Swiss Department of Agriculture, began a joint project with the Swiss Farmers' Secretariat to establish good practice in farm accounting among Swiss farmers. Laur was extending the work of his German contemporary the economist Freidrich Aereboe, widely regarded as a great innovator in farm practice, and that of other economists working on cost of production methods in agriculture. Initially, a small group of farmers attended bookkeeping classes and were then paid to keep these books for one year and to contribute their accounts to a pool kept by the Secretariat. The programme was sponsored by the government; by 1901 there were 131 farmers involved and by 1926–27 the number was 487 single-entry systems and five double-entry systems. A number of farmers continued to keep the records after one year, with about 11 per cent keeping up the records for 10 years or more. The results were published in an annual yearbook. In 1927 Hinrich reports that 8–20 people at the Secretariat were required to close off the accounts from the books provided by the farmers. The annual statements were then used as the basis of statistical analyses. There were debates at this time over the usefulness of cost-of-production studies, which is how Laur characterized his work, but he defended his work by 'insisting that agricultural economics could render no greater service to agriculture than actually to determine the true cost of production' (Hinrichs, 1929: 651).

Hinrich suggests that Swiss farm accounts under the system established by Laur contributed:

- a determination of Swiss agricultural incomes and their changes under the influences of natural and economic factors;
- solutions to numerous farm management problems;
- a supply of statistical materials to farmers and agricultural schools for educational purposes;
- evidence supporting demands for agricultural legislation;
- a dependable basis for land valuation, rental and inheritance purposes.

All of which are still cited as the benefits of comparative systems today.

Similar schemes to collect together farm accounting data were set up independently in Czechoslovakia (now the Czech Republic and Slovenia), Denmark, France and the Netherlands during the early part of the twentieth century. There were variations on the idea – in Denmark, a peripatetic bookkeeper did a monthly round of farms to bring the accounts up to date and to collect the data on behalf of the economists running the scheme – but essentially the basis was the same. Farm accounts were collated and a comparative analysis was prepared, with information being fed back to farmers but also providing a detailed resource for agricultural economists.

Collecting Data by Survey Methods

The survey method was favoured in the US, where the first attempts were made in the 1890s by G.F. Warren (a prolific writer on farm management in the early twentieth century), who surveyed orchards in Western New York using a statistical approach. It is reported that 'the financial and farm-management phases of the situation were reported with care' along with geological factors. G.F. Warren went on in 1909 to publish 'The income of 178 New York Farms' and other surveys began to follow suit. The ethos behind the early surveys was that 'every farm is an experimental station and every farmer a director thereof' (Warren, S.W., 1945). The ideal was to learn from the collective experience of the experiences and experiments of all farmers.

The survey method was also seen as a means of getting information from illiterate or semi-literate farmers. 'It means getting data by personal visits to farmers. A farmer may be able to answer questions from memory or from his records or both' (ibid. 20). By the 1940s, there were over 40 surveys in operation in the US and the idea was taken up in England in 1936, in the form of the farm business survey. S.W. Warren (1945: 23) summarized the problems that arise in using farm data in this way, which remains a criticism of mass participation benchmarking:

> Our big need for the future is to obtain groups of farm management records which are homogeneous, not only as to soil, climate, topography and markets, but also as to the education of the operator, acreage of the farm, and some other important factors. If we can eliminate these major factors by the sampling process we can then go ahead to study some of the minor factors affecting farmers' incomes.

The statistical approach to identifying best practice can lose the detail that makes benchmarking using small group or one-to-one methods more effective in identifying specific operational changes that can be made.

To an extent, the work in agriculture was indicative of developments in other industries. In the period 1900–25, an interesting phenomenon was observed in the US. As identified by Berk and Schneiberg (2005) a number of sectors were trying something different: 'printers, bridge builders and foundry men, among others, turned from co-operative price controls to collaborative efforts to enhance productivity' by 'designing associations to elevate rather than suppress competition through deliberation, benchmarking and learning'. In essence, firms otherwise in competition with each other pooled accounting and production data in order to 'craft standard accounting systems', which were used to combat the pricing under cost and oversupply which threatened the competitiveness of all firms in a sector. This was largely done through setting industry cost averages as benchmarks which were then disseminated through the industrial trade associations. Berk and Schneiberg (2005: 49) state, 'By 1925, at least 25 percent of all American trade associations had participated in crafting this alternative, with up to half that number instituting the logic of developmental association in whole or in part.' One of the early adopters of this form of benchmarking was the National Cooperative Milk Producers Federation (ibid. 70).

The innovation of collecting farm accounting data by survey or by pooling of data gradually diffused through agriculture. The 'Finbin' system visited in the project reported in Chapter 3 began in 1929 as a farm extension project. Everett M. Rogers, whose work on the diffusion of innovation had its roots in his studies of the spread of agricultural practices, comments on the agricultural extension service in the US, which he says is 'reported to be the world's most successful change agency'. The agricultural extension model is 'a set of assumptions, principles and organizational structures for diffusing the results of agricultural research to farmers in the United States' (Rogers, 1995: 357). These structures included the US Department of Agriculture (USDA) and land grant universities established by Congressional legislation passed in 1862 (ibid.). More importantly for benchmarking, the use of county extension agents and a farm bureau by Broome County, New York in 1911 set the format still in operation today (ibid. 359).

Assisted by the University of Minnesota and USDA, and working through the farm bureau, the *First Report of the Better Farming Club of Waseca County* was published in 1929. The foreword states that 'The main purpose of the club is to secure such data and information, which when compared with that secured on other farms will enable the co-operator to increase his efficiency in various enterprises and to organise the farm on a more profitable basis'. Records kept included 'inventories at the beginning and end of the year, cash receipts and expenses, a report of feed fed to various classes of livestock, and a record of farm produce used by the farm family'. The record keeping was assisted by the field agricultural extension worker who supervised the system and added information as needed, with a representative from the University of Minnesota checking for completeness and accuracy. Data was collated and reported in the report, along with returns for labour and investment. The different tables presented included:

- index of operating earnings;
- pounds (lbs) of butterfat per cow;
- return per $100 feed;
- number of cow units per 100 acres;
- pounds (lbs) of pork per feed crop acre;
- number of productive livestock units per 100 acres;

- eggs per hen.

The report comments, 'It is quite evident from this report that few farms have a monopoly on efficiency.'

– 11 –

Thermometer Chart

Using your figures from page 10, locate your standing with respect to the various measures of farm organization and management efficiency. The averages for 131 farms included in this summary are located between the dotted lines across the center of this page.

Oper. labor earnings	Crop yields	High return crops	Return from productive livestock	PR. L.S. units per 100 acres	Work units	Work units per worker	Pow. mach. e.g. &bldgs etc. per unit
$9600	132	82.0	140	35.5	765	405	405
8800	128	79.0	135	33.5	730	390	390
8000	124	76.0	130	31.5	695	375	375
7200	120	73.0	125	29.5	660	360	360
6400	116	70.0	120	27.5	625	345	345
5600	112	67.0	115	25.5	590	330	330
4800	108	64.0	110	23.5	555	315	315
4000	104	61.0	105	21.5	520	300	300
3200	100	58.0	100	19.5	485	285	285
2400	96	55.0	95	17.5	450	270	270
1600	92	52.0	90	15.5	415	255	255
800	88	49.0	85	13.5	380	240	240
0	84	46.0	80	11.5	345	225	225
-800	80	43.0	75	9.5	310	210	210
-1600	76	40.0	70	7.5	275	195	195
-2400	72	37.0	60	5.5	240	180	180

Figure 2.1   Temperature graph – mimeograph from farm report of southeastern Minnesota, 1949

Source: University of Minnesota, 1949.

In 1929, 27 farms averaging 157 acres each co-operated in the club. Figure 2.1 shows an early form of comparing farm performance as a 'temperature' against an average derived from all farms in the database. The project soon rolled out across South West Minnesota and reports were expanded to include more detailed enterprise reports, retaining a similar format until 1976, when a new team took over and computerization was introduced. This allowed for whole farm reports to be produced and new measures to be calculated, which in time came to include those measures recommended by the Financial Standards for Farming, which are discussed later in the chapter. Data is collected anonymously through the Farm Bureaux in the State and filtered reports are fed back through the Bureaux. The club that began in 1929 grew into the Finbin system identified in Chapter 3 as one of the best in class for mass participation benchmarking.

National Surveys

Surveys were conducted in Britain from the mid-1920s. James Wyllie, a researcher from Wye Agricultural College in Kent, England (now part of Imperial College University), reported his first survey findings in 1925, and in 1934 the Agricultural Economics Research Institute in Oxford established the National Milk Costs Investigation Scheme. The University of Cambridge set up the Cambridge Food Recording Scheme in 1936, the same year in which the Farm Management Survey (now the Farm Business Survey) was established. American economist Warren's 1929 publication of his experiences conducting surveys in the US was influential on the development of these latter schemes. By the post-war establishment of the Provincial Agricultural Economics Service the Farm Management Survey had been rolled out across the country, co-ordinated by the Ministry for Agriculture, Fisheries and Food (MAFF) but carried out in the universities-based Provincial Agricultural Economics Service which was in place at the time. This survey is still performed by Agricultural Economics Units in key universities, including Nottingham, Exeter, Reading, Cambridge, Newcastle and Imperial College London, and managed by the Department for Environment, Food and Rural Affairs (Defra), but is now referred to as the Farm Business Survey. The findings of the survey are published annually as the publication *Agricultural Incomes in the United Kingdom*. The Cambridge Food Recording Scheme set the model that the Farm Business Survey follows, by being 'a comparison of the individual farmer's results, factor by factor with sample averages, best and worst groups and for certain factors with established theoretical standards' (Lloyd, 1970). By the late 1930s, the specific objective of using the Farm Management Survey to provide farm management standards was formed and, as we shall see, became central to the post-war farm management policy.

The Farm Business Survey in the UK was one of the prototypes of a much larger collection of survey data set up in 1965 through the European Economic Community (now European Union) Common Agricultural Policy. This is the Farm Accountancy Data Network (FADN). At the European level, the farm account surveys of all the Member States are brought together under the co-ordination of the agricultural Directorate-General of the European Commission (DGVI) for the FADN, also known by its French acronym RICA. This was established in 1965 'with the specific objective of obtaining data enabling income changes in the various classes of agricultural holding to be properly monitored' (Hill, 2000). The justification for FADN was rooted in policy, in that 'the development of the Common Agricultural Policy requires that there should be available objective and relevant information on incomes in the various categories of agricultural holdings and on the business operation of holdings coming within categories which call for special attention at Community level' (EEC Regulation 79/65). FADN is therefore not a single survey but an amalgamation of national surveys carried out by Member States. (Hill, 2000: 156) Based in the Netherlands, this significant database draws on established systems of data collection, such as the FBS and the French CER survey, and the tax-based Danish system, but also required the establishment of new surveys in certain countries, such as the LEI in the Netherlands. Altogether, the FADN sample consists of about 60,000 holdings (EUR 15) corresponding to just over one per cent of all holdings within the FADN's field of observation. The sample is stratified by economic size (in European Size Units), by farming type and region, and all published results are weighted appropriately. In the strictest sense, the sample is not random, since it is drawn from holdings that keep accounts, and participation by farmers is voluntary. The database is used primarily by researchers and policy informers, and although it has been suggested that further use could be made for benchmarking purposes by farm businesses (Argiles and Slof, 2001), there is no extensive evidence that it is being used at this micro level.

The results of the surveys were not published for use by farmers until the 1960s. Lloyd (1970), commenting on the original surveys in the 1920s and 1930s, says that there was 'no organized system for disseminating the results to farmers in a form that could assist them with their farm management. Neither was there a very penetrating analysis made of causative factors in the economy of the farm business.' He goes on to say, in relation to the fact that the data was used to inform economists and advisors, that 'in defence of the methods used, it has been suggested that the agricultural economics departments safeguarded the sensible business use of accounts and costings by sending into the advisory field only experienced personnel. However, it has also been commented that such personnel represented more an insurance against accounting errors than a means of giving effective advice'. The reliability (or in some cases existence) of farm record keeping and the accuracy of the data being collected has always been an issue in farm surveys.

Although both the FBS and FADN have developed robust methodologies for the collection of data, and are now recognized as providing reliable statistical data for analysis, they are both still voluntary schemes. There is a limit on how long farmers can remain in the FBS (10 years with a gap of 10 years before they can rejoin), but there is still the criticism that the sample is skewed towards those farmers who are more progressive and keep records, rather than being a truly representative national sample.

Another issue in relation to national surveys is just how useful the highly aggregated data is for providing benchmarks for individual farm decision-making. The University Centres that collect the data for the FBS publish regional figures and the computerization of the survey allows filtering and 'drilling down' of data. Defra launched an online comparative analysis service called Farm Business Survey Benchmarking Online in 2008 and the LEI in the Netherlands offers a similar product. The Australian statistics offered through the Australian Bureau of Agricultural and Resource Economics (ABARE) can be interrogated online (www.abare.gov.au).

A report commissioned by the Food Chain Centre in the UK which examined a range of benchmarking schemes found that 'LEI collect data which the Dutch government require for EU purposes, benchmarking is an add-on designed to deliver value back to the participating farmers (*a quid pro quo*). This prioritizes data collection for the government not the delivery of value to farmers'; and as a consequence, the report says, there was less trust and reliance on this source of benchmarking data. However, in the UK participants generally welcome seeing how they compare to other farmers, and regret the periods when they are ineligible for the scheme. The extent to which they make decisions based on the data is, however, limited – for many it is a matter of assurance or at best, an indicator to take some course of action – but the main value of the data is still to advisors.

Introducing Ratios and Indices as Benchmarks – The Development of Comparative Analysis in the UK

During the Second World War, the need to decrease reliance on imported foodstuffs and increase food security was a driver both in Europe and the US for putting policies in place that would increase yields and productivity in farming. This meant investing both in technology and in improving the management skills of farmers. In Britain, one of the provisions of the Agricultural Act of 1947 (of which one promise was to deliver profitability to farmers) was the setting up of the National Agricultural Advisory Service (NAAS). Advisors were trained and employed by NAAS to go into farms to help farmers diagnose problems on the farm and to suggest solutions. These advisors were supported by agricultural economists based in universities around the UK, in what was termed the Provisional Agricultural Economics Service or PAES. Previously, advisory work on the farm had been undertaken by the same economists, as we have already seen. When the relationship between the two groups proved difficult to negotiate, MAFF appointed Farm Management Liaison Officers (FMLOs) with the express task of developing analytical tools with the economists in the PAES that could be used on the ground by the advisors in NAAS. The prime mover in this task was one Charles Blagburn, the FMLO at the University of Reading.

Blagburn and fellow FMLOs realized that they had two sources of data on hand. The first was the survey data being collected by the PAES for the Farm Business Survey (then called the Farm Management Survey) begun in 1936. Second, in 1941 farmers in Britain had become subject to income tax for the first time (previously they were taxed according to land rental values) and as a consequence, farmers were now obliged to have annual fiscal accounts for their farms. The advisors would collect the necessary data on farm and record it on paper spreadsheets and forms which would then be sent to the PAES for analysis. By analysing both sets of data using ratios, a set of comparative figures could be generated which would be used by the advisors to diagnose weaknesses on individual farms. Given that the PAES were obliged to carry out the FMS work, a considerable body favoured further standardization of accounting and costing techniques and 'the development of a system of business efficiency measurement which could use the data accumulated as a by-product of the Ministry's required work programmes' according to Lloyd (1970). A set of 'yardsticks' (or benchmarks) were developed based on an American idea called 'system and yield indices'. This was known as 'whole farm comparative economic analysis'.

This system was used throughout the 1950s in the UK but had one practical fault: the paper systems were cumbersome and took significant amounts of time to generate. By the end of the 1950s, a computerized system was being developed to facilitate the production of more timely information, but other issues had arisen. Government assistance had been successful in increasing the yields and the incomes of farmers, through price support mechanisms. The time was now right to consider marginal improvements in both yields and incomes that could be effected by the management skills of the farmer rather than simply by improved technology. An FMLO in Northern Ireland, Liversage, proposed a system of enterprise costing and planning based on gross profits – that is, taking the variable costs from outputs (at fixed prices) as a tool for finding the optimal mix of enterprises on the farm. Aggregate gross profits per enterprise less fixed overheads would give the net profit. The idea was championed by another FMLO – David Wallace of Cambridge University – and renamed 'gross margins'. Wallace developed the gross margin analysis and planning system and promoted it through the BBC and local meetings (Jack, 2005).

By 1972, Gross Margin Accounting in UK agriculture was 'nearly universal', according to John Nix, a professor at Wye College who had begun his career as one of the 'second generation' FMLOs. Nix himself is famous for having developed a handbook of enterprise gross margins in 1966 which was designed to assist farmers with their budgeting practices. It was definitely not a tool for comparative analysis, but within a very short period of time that is what it was being used for in the field, 'the least safe use of gross margins', as Tony Giles later commented (Giles, 1986).

Comparative analysis was well established in Britain by the 1970s. On the one hand, there was the Farm Business Survey, which provided comparative data based on Net Farm Income calculations. This underwent some modifications along with the FADN requirements but was being run on essentially the same principles in the early twenty-first century. On the other hand, a significant number of computerized 'benchmarking' systems had been developed by levy boards such as the Red Meat Industry Forum and the Meat Livestock Commission, by private consultants (a growing profession from about 1960), computer software companies and trade associations. Initially, in the 1960s, farmers posted their accounting data to the provider who then processed the results on their computer and made returns to the farmer. Both ICI and Pauls BOCM ran this as a service for loyal clients in the 1960s. Nowadays, the majority of systems allow online data submission, either by advisors or by farmers themselves. These systems in the UK are almost entirely based on gross margin formats, and a division of costs into variable and fixed. MAFF (now Defra) also published special reports on individual sectors and enterprises based on gross margins. There are a small number of more recent systems that use a cost per unit basis, notably in the Dairy Industry (such as the ADAS/HBSC Spotlight figures).

The Australian Experience and the Critics of Comparative Analysis

Australia was a later arrival to the widespread use of comparative accounting analysis for agriculture. By the early 1960s, the Queensland Department of Primary Industries and the University of New England in New South Wales had mail-in comparative analysis programmes. There had been a growing trend towards criticizing the comparative accounting models of other countries by a number of Australian academics and practitioners, but there was another faction who supported the introduction of farm accounting groups and State-led comparative analysis programmes. Because farm accounting was led by the production of financial accounts for tax purposes, there was a call to standardize farm accounting. The idea was to have a standard format for financial accounts that would be used by all accounting firms, with farm recording being done against a standard chart of accounts. The accounts would include a profit statement, a statement of assets and liabilities, and a statement of sources and uses of cash, whilst the bookkeeping would be facilitated by cheque books and books of entry that allowed transaction details to be coded up according to the chart of accounts. There was a similar project in New Zealand from 1961.

The Australian project – the Joint Committee on the Standardization of Farm Management Accounting – produced a report called *Accounting and Planning for Farm Management* (better known as 'The Blue Book'). It also produced the Australian Chart and Code for Rural Accounting (ACCRA), but take-up was patchy. Although the principle of having one set of accounts that would feed into a central database for analysis purposes was logical, there were many theoretical and practical issues. Malcolm (1990) comments that there are three problems with ACCRA. Firstly, there was a lack of interest on the part of farmers in record keeping. Secondly, there was accountants' predilection for concentrating almost exclusively on tax accounting. Thirdly, there was the existence of alternative competing accounting systems. The potential for accountants to add value to accounting services by offering a more bespoke set of comparative analysis figures outweighed the advantages of having a national system. More theoretically, historic financial accounting data is an unsuitable basis for making future management decisions and the reconciliations required to validate the statements of account were too obtuse for most farmers.

Makeham and Malcolm (1993: 552) summed up the arguments against comparative accounting by saying:

> *A major fallacy was the belief that historical records, and comparative analysis of technical ratios and average activity gross margins, achieved on different farms were useful for farm management analysis. They are not very useful. Farm management is about dealing with what might happen. What happened in different, past circumstances is of limited relevance. The weaknesses of emphasis on accounting and recording is that generally it leaves out of the analyses the critical technical and human aspects and the management economics way of thinking, and has a past, not future, orientation.*

However, there was one strong point in the thinking of the Australian and New Zealand academics and practitioners developing these standardized systems. Practising accountants in these countries have had much more influence on the development of farm accounting than in other countries. Part of this is due to historical accident (Carnegie and Napier, 2002), in that the accountants who had emigrated to the New World in the nineteenth century realized that either they specialize in farming or mining, or change profession. The need to service clients meant that more rigorous accounting methodologies – double-entry bookkeeping and ledgers rather than single-entry cashbooks, for example – were developed in the UK and the US, and accountants in practice took more interest in the use of comparative analysis. Therefore, as most farmers would use an accountant to prepare the accounts, it would make sense for them to enter that same data into a database for the production of benchmarks. This same principle is being used in the DairyBase system being pioneered in New Zealand (the subject of the study in Chapter 4).

Despite the criticisms and practical difficulties surrounding comparative analysis in Australia (Fleming et al., 2006; Ronan and Cleary, 2000), it has become institutionalized. A paper from Australia by Worsley and Gardner (2000) lists 66 different databases run by consultancy firms and trade associations, and each State compiles aggregate statistical data for analysis based on economic and gross margin figures.

The US and the Sweet Sixteen

Farm management academics and advisors in the US moved more towards budgeting and linear programming approaches to solving farm problems in the 1950s and 1960s, and mass participation comparative analysis continued along the lines above. Towards the end of the 1980s, however, the desirability of having consistency between the different data sets and the advice given led to the Farm Financial Standards Council in the US issuing recommendations to guide financial analysis of farm data (Fedie, 1997). This set out a standard set of accounts and ratios generated from them. Unlike whole farm comparative economic analysis, these are financial accounting ratios, similar to many used to assess the health of any business. These are known colloquially as the 'sweet sixteen', and cover liquidity, solvency, profitability, repayment capacity and financial efficiency. They are widely used in the outputs of mass participation benchmarking systems in the US and by farm advisors.

The Characteristics of Successful Benchmark Providers

The Food Chain Centre in the UK commissioned a report in 2006 which in effect benchmarked the benchmark providers in farming and food. The authors of the report examined 14 'mass participation' benchmarking schemes in North America, South Africa, Australasia, Ireland, Denmark, the Netherlands, France and Belgium. Each scheme had between 300 and 3,000 participants. The providers of benchmarking services (all comparative analysis schemes) were clustered as follows:

1. Development initiative from advisory service/levy board/trade association.
2. Governmental statistics provision oriented.
3. Piggybacked on farm software system.
4. Piggybacked on lender requirements.

Successful schemes – that is the ones with a high level of participation and ongoing use – were found to be provided by an advisor network. Advisors both promoted the database and gave feedback and interpretation of the results. Such schemes also prioritize farmers and their information needs. Another factor in participation was where there was an influencer, such as the programme being offered as a service by a bank (which in turn could use the data to assess lending decisions); the provision of accounting services; the connection to a trade association or board to which the farmer belonged; or the link to supply chain partners. A clear message that the information would assist in decision-making and by implication improvements in financial performance was also an inducement to join. Anonymity was an important factor for some participants, as were the efficiency of the service, web delivery and the extent to which data could be filtered and drilled to provide like-for-like data. The details of this study are found in Chapter 3.

One of the key reasons for mass participation, as perceived by the providers, was that a low level of effort was required in order to participate and benefit from benchmarking in this manner. Another was the demonstration of performance relative to other farm enterprises. Another highly ranked attribute was that financial improvements were made as a result of opportunities and weaknesses highlighted, but again there is a lack of concrete evidence to prove this.

Farmers that use advisors and others to assist with the recording of data are perceived to be more successful than those that either do not keep records or do not use an advisor. The strength of comparative analysis is to advise the *advisors* of farmers, rather than the *individual farmer*. There are criticisms of the methodology of comparative analysis, but an early response to this still holds: 'most of the criticisms of the use of comparative efficiency measures cannot be levelled at those who have carried out a full analysis and built a balanced interpretation from a group of yardsticks with which they are familiar' (Lloyd, 1970). But the potential for benchmarking lies with those farmers who 'know their costs' and participate more fully in the process of benchmarking rather than act as more passive recipients. In the next chapter, the rise of discussion groups and more individualized benchmarking for best practice in farming is examined, showing that comparative analysis and the provision of benchmarks alone are just the first step in unlocking the potential of benchmarking in the industry.

International Benchmark Data

Besides the Farm Accountancy Data Network shown above, the International Farm Comparison Network has produced figures for international comparisons for the dairy sector. This is a high-level analysis and so less relevant for the analysis of individual farms, but could be an indicator of which countries might provide farms against which 'competitive' benchmarking could be carried out and this suggestion is explored in Chapter 9.

Strengths and Weaknesses of Comparative Analysis

The general problems with comparative analysis have been recognized from an early stage. Lloyd (1970) commented that it was 'not good for a-typical farms; high performers (already in or above premium group); simple farms, because of the danger of over-analysis; complex farms, because of the dangers of compensating effects; situations where there are rapidly narrowing unit margins, due to falling product prices or rising resource costs'. Longworth and Menz (1980) note that farming is about 'moving towards a continually improving and adjusting efficiency', whereas economic models such as those on which performance measures in agriculture have tended to be based assume a level of technical efficiency.

In the 1950s, the conditions in the UK were suited to comparative analysis. It was observed that whole farm comparative economic analysis was particularly good for farms with below average performance, yet that is only to be expected: as Sturrock said in his preface to Selly and Wallace (1961), 'one sometimes hears the comment that farm analysis is only for the "below average" farmer. In fact the badly run farm hardly needs analysis at all – the faults are so obvious that they can be recognized on sight'. An advisor is likely to have considerable success the first time that a farm business seeks advice, and particularly in the immediate post-war period, when stocking levels and intensity were artificially low, the gains from increasing output could look disproportionately favourable. Dissatisfaction arises with comparative analysis in the difficulties in diagnosing the problems on high performance and revisited farms. Even with gross margin analysis, Giles and James (1993: 236) report the following comments by David Wallace:

> *Looking back, he reflects on the fact that at the time 'making big improvements in farm income was easy, given a reasonable level of technical competence on the farmer's part. Simplification, bigness and new technology all meant that the sums were easy – and (he continues) I was glad to be out of it by the time real problems of choice and overproduction created a very different climate.'*

Therefore, given that the most reliable records feeding into benchmarks in mass participation schemes are from the top performing farms, the farmers best equipped by education and experience to make sustainable changes on the basis of management information belong to the groups not amenable to comparative analysis. The benchmark figures can act as a reassurance or indicator of the right direction, that the farmer is keeping up with, or ahead of, the neighbours, and provide a validating figure, but otherwise, as one farmer said of the scheme they belonged to, 'it's just an interesting benchmark'. It may stimulate interest but it is not evident that it might stimulate action that leads to change.

The problem with the use of benchmarks derived from aggregating and averaging financial accounting data is that there is very little evidence that the availability and use of this data has any impact on farm decision-making through changes made. There have simply been no published studies on how farmers use this information for decision making. One study from the US into the management practices of a sample of dairy farmers indicated that 'While the [benchmark] practices are widely adopted, their adoption does not appear to influence profitability in this sample of farms' (Gloy and LaDue, 2003: 169). They also found that the top performing five per cent of farmers in the programme examined would not think of making a decision without considering the benchmark data provided by the scheme, but this was in the view of the operators of the scheme, and not a sample of users.

Fleming et al. (2006) identify five key problems with comparative analysis from the academic literature:

1. it failed to incorporate sound economic principles in its application;
2. there was limited scope for action once indices were calculated;
3. the approach failed to establish causal relations between farming practices and performance;
4. it was not consistent with a holistic approach to farm decision-making;
5. risks and uncertainty in farm decision-making were neglected.

They follow Malcolm's (2004) scathing comments on the resources spent in Australia in establishing comparative analysis systems and adopt his comments on the lack of economic discipline to provide a framework for rational choice based on data collected by saying:

> Lambasted research and development organisations in Australia that 'have invested substantial funds in conducting large scale "average benchmarking" or comparative analysis studies with on-farm diagnostic and prescriptive intent'. State departments of agriculture were also the targets of his criticism, in that they 'have invested large amounts of resources over long periods of time conducting comparative analysis for farm management' with little payoff.

> (Malcolm, 2004: 396)

Fleming et al. (2006) attempt to address each of the five points, and themselves suggest that small group benchmarking, discussed in Chapter 5, overcomes the identified problems and they posit five beneficial features of small group benchmarking, namely that:

- farmers who pay to belong to a group have a vested interest in providing accurate data;
- facilitators are familiar with day-to-day farming operations, which survey data collectors may not be;
- consultants can then offer feedback to those compiling larger scale data sets;
- following that, modelling becomes closer to actual results and practices;
- consultants and facilitators with their groups can discuss issues of risk and uncertainty that are not accounted for in traditional comparative analysis outputs.

The limitations of comparative analysis and mass participation benchmarking to help high-performing, individual farmers can be overcome by more focused methods such as small group work and process benchmarking, as discussed in later chapters. An advisor in the East of England, who ran computer courses for farmers where they compared their own data to the averages in the Farm Business Survey as part of an exercise, made the following comment:

> I think where that whole operation falls down is the fact that it is just a snapshot and the farmer goes away and it's up to the farmer to do something. I very much feel that we've got to a stage surely where we can put that into a programme, so that the farmer does that and then somehow, he is then taken along a certain track, so that if he's incredibly good then what I want him to do, I want him to go off into a master class. I want him to be going into a group of really turned on and thoroughly interesting individuals and he'd just get so much of a buzz out of just being in with an entrepreneurial crew that they would just enjoy their own company. Then, the middle guy, who is just looking for a bit of direction say, he ought to get into business plans, he wants to make sure he's got a real business plan which he updates annually, perhaps a five year plan, updating it annually, and that's all that we want him to do because he then starts to think about where his business is off to.

> (As told to the author)

What then is the strength of setting benchmarks in this way, through mass collection of largely financial data, presented as quartiles of high, medium and low performance? The existence of hundreds of such schemes developed over a significant time period suggests that they have value. They have value in providing diagnostic data and trends that help the advisor frame their advice in terms of whether a downward turn has happened to everyone or is due to a farmer's own actions. It is also widely believed – or at least hoped – among advisors that the trends and rankings that emerge from the presented data will motivate farmers to change practices and to seek help. The mini case study in the box illustrates the thinking behind this use of the data, and gives evidence about how this data is used. As becomes clear, it does not give evidence about the decisions taken on the basis of the data nor on the long-term changes that were brought about. The author, indeed, has not been able, so far, to find any such evidence – benchmarks and indicators from mass participation tend to be used to confirm that what has happened was satisfactory or to diagnose why it might not have been satisfactory, or to provide figures for business plans, but with regards to sustainable change, a definitive study is still outstanding.

MINI CASE STUDY: EAST OF ENGLAND FARM ADVISOR

An advisor in the East of England, who also still owns a farm on which the work is now contracted out, explains how he uses benchmark data in his own business and in his advisory team.

'Using the Farm Business Survey, I used to rank us on the gross margins of each of the crops, compared to other people. I used to rank us on efficiency of labour and machinery and then come all the way down to the bottom to Net Farm Income (NFI) and that's what really interested me … because that's the amount of money I can put back into the business – is it worthwhile being in business? Interestingly, in [the latest Eastern region survey], the NFI for the middle of Suffolk is negative this year but for Norfolk it's still quite a high positive figure, and having read that survey over the weekend, I was going to punch it round to my team, because each of the six of them, the different areas the six of them are running are huge differences between the NFI between various counties.

It just seems that the "heavy lands" that go through the middle of Suffolk and up towards Norwich, then comes down and swings off through the top of Essex and out towards Bedfordshire, that really suffered from having a bad Autumn, what, 18 months ago so the winter wheat that that land relies on never really got away. Whereas the lighter lands up near Norfolk and so on seem to have had a good season … there's actually agronomic, climate reasons why that particular part of the region didn't do so well … [What] I was trying to pick out was, is it just gross margins, is it the crops they grow, and it seems [as] though the variety of the crops differ so greatly … In the summary report they were picking out particular reasons why the parts of the region didn't do so well, and one [reason] was the weather. But what I was trying to illustrate to the boys is that you can go onto a Norfolk farm and he's still made money last year, you can go onto a Suffolk farm and he's lost, and that's going to make a difference to the attitude of the farmer, so that the Suffolk guy should go in there and … spend a bit more time reassuring the farmer … And whereas the farmer who's just seen his NFI drop from £500 to £400, he's starting to see the trends but the farmer who's dropped from sort of £500 to nought, or minus 20 or whatever he's thinking "Oh Boy! Got to do something quickly" and he has a different attitude towards developing something on his farm.

'If you ask me personally, I picked up [four years ago] that I couldn't read any parameters that said that the value of what I produced on the farm was going to go up. And over the last four years it's just tracked down. I always write an annual review at the end of each year, so that I can look back [at that first year] and say "well what triggered me to start to get very focussed on the viability of my farm?" And ... what interested me about my actions, if I look back on it, one of the first things I did was to protect the income that I was used to, and I also wanted to gain extra skills in order to be able to do other things away from the farm ...'

(As told to the author)

Benchmarks for Comparing Farm Performance and Sustainable Change

As seen in Chapter 1, the benefits of benchmarking are that it can:

- create motivation for change;
- provide a vision for what an organization can look like after change;
- provide data, evidence and success stories for inspiring change;
- identify best practices for how to manage change;
- create a baseline or yardstick by which to evaluate the impact of earlier changes.

Despite the lack of non-anecdotal evidence, it could be said that the availability of these benchmarks for comparative performance might create motivation for change among those ranked in the middle or lower quartiles. However, it is not unusual for those in high-performing farms to say that they do not benchmark, because their farm is not like anyone else's farm – they would 'skew the average' and get no useful information. Similarly, figures may be skewed because poor performers simply do not volunteer to participate: databases are largely self-selecting. The potential for creating motivation for change and providing evidence of the need to change is there, but the problem is that due to the anonymous and aggregated nature of the figures, and the lack of information about individual operations, it is unlikely that best practices can be identified through this data: this is one of the biggest weaknesses of comparative analysis in this form, in regard to it being properly utilized as benchmarking, as the term is understood conventionally. Similarly, the only vision provided is a financial one, which on its own may not be inspiring: many studies have shown that generating or maximizing profits alone are rarely the key objective of farmers. However, there is a baseline against which to evaluate change, at least in financial terms.

It is almost impossible to comment on whether benchmarks in this form would drive innovation. Certainly, there is innovation in creating and presenting the benchmark systems themselves – as Chapters 3 and 4 show amply – and they have driven innovation in providing advisory and analytical services to the farming industry. But the evidence of the impact on farming practice is lacking. The figures produced also impact on policy making for the industry – that discussion, however, is outside the scope of this book.

A final comment is that the existence of such mass databases has driven research and learning in the industry, for advisors, academics and for individual farmers. This in turn has contributed to the innovation explored in Part 2, Benchmarking for Better Practice. The development of small group benchmarking in agriculture is a true innovation and in turn has led to adaptive innovation in farm business and operational practice, but its roots can be found in the databases based on farm accounts that have grown since the end of the nineteenth century.

References

Argiles, J.M. and Slof, E.J. (2001), New opportunities for farm accounting, *The European Accounting Review* 2001 Vol. 10 No. 2, 361–83.

Berk and Schneiberg (2005), Varieties in capitalism, varieties of association: collaborative learning in American Industry 1900–1925, *Politics and Society* Vol. 33 No. 1, 46–87.

Carnegie, G.D. and Napier, C.J. (2002), Exploring comparative international accounting history, *Accounting, Auditing & Accountability Journal* Vol. 15 No. 5, 689–718.

Fedie, D.M. (1997), *How to Farm for Profit: Practical Enterprise Analysis* (Iowa: Iowa State University Press).

Fleming, E., Farrell, T., Villano, R. and Fleming, P. (2006), Is farm benchmarking the new acceptable face of comparative analysis?, *Australasian Agribusiness Review* Vol. 14, Paper 12.

Giles, A.K. (1986), Net margins and all that – an essay in management accounting in agriculture, reprinted in *Windows on Agricultural Economics and Farm Management (1993)* (Reading: College of Estate Management, University of Reading).

Giles, A.K. and James, P.J. (1993), Farm management liaison – a unique era, reprinted in *Windows on Agricultural Economics and Farm Management (1993)* (Reading: College of Estate Management, University of Reading).

Gloy, B.A. and LaDue, E.L. (2003), Financial management practices and farm profitability, *Agriculture Finance Review* Vol. 63 No. 2, 157–74.

Hinrichs, A.F. (1929), Swiss studies in farm accounting, *Journal of Farm Economics* Vol. 4, 648–51.

Hill, B. (2000), *Farm Incomes, Wealth and Agricultural Policy.* 3rd edition (Aldershot: Ashgate).

Jack, L. (2005), Stocks of knowledge, simplification and unintended consequences: the persistence of post-war accounting practices in UK agriculture, *Management Accounting Research* Vol. 16, No. 1, 59–79.

Juchau, R. and Hill, P. (2000), *Agricultural Accounting: Perspectives and Issues.* 2nd edition (Wye: University of London, Wye College).

Lloyd, D.H. (1970), *The Development of Farm Business Analysis and Planning in Britain: A Methodological Review and Appraisal* (Reading: University of Reading Dept of Agriculture, Study No. 6).

Longworth, J.W. and Menz, K.M. (1980), Activity analysis: bridging the gap between production economics theory and practical farm management procedures, *Review of Marketing and Agricultural Economics* Vol. 48 No. 1, 7–20.

Makeham and Malcolm (1993), *The Farming Game Now* (Cambridge: Cambridge University Press).

Rogers, E.M. (1995), *Diffusion of Innovations.* 4th edition (New York: The Free Press).

Malcolm, B. (2004), Where's the economics? – the core discipline of farm management has gone missing, *Australian Journal of Agricultural and Resource Economics* Vol. 8 No. 3, 395–417.

Malcolm, L.R. (1990), Fifty years of farm management in Australia: survey and review, *Review of Marketing and Agricultural Economics* Vol. 58 No. 1, 24–55 in Juchau, R. and Hill, P. (eds.) (2000), *Agricultural Accounting: Perspectives and Issues*, 2nd edition. (Wye: University of London, Wye College).

Ronan, G. and Cleary, G. (2000), Best benchmarking practice in Australian agriculture: issues and challenges *Agribusiness Perspectives* – Paper 39 <http://www.agrifood.info/Review/Perspectives/2000_Ronan/2000Ronan>.

Scorgie, M.E. (1997), Progenitors of modern management accounting concepts and mensurations in pre-industrial England, *Accounting, Business and Financial History* Vol. 7 No. 1, 31–59.

Selly, C. and Wallace D. B. (1961), *British Broadcasting Corporation: Planning for Profit: A Simple Costings System to Help Farmers Meet the Challenge of Falling Prices* (London: BBC).

Warren, S.W. (1945), Forty years of farm management surveys, *Journal of Farm Economics* Vol. 27 No. 1, 18–23.

Worsley, A. and Gardner, M. (2000), *The Short Report: No 74: Rural Benchmarking Programs – A Review*, (Australia: Rural Industries Research and Development Council) <http://www.rirdc.gov.au/pub/shortreps/sr74.htm>.

3 Best Practice in Mass Participation Benchmarking

SARAH THELWALL
Prospect Management Services

Mass participation by farmers in benchmarking schemes requires that participating farmers achieve specific and measurable benefits. The benefits which are most highly valued are those which contribute to increased profitability of the farm enterprise and increased competitiveness in comparison with their local, regional and international competitors. Additionally, there are a number of other benefits to benchmarking which enhance its overall value to the agricultural supply chain at a variety of points in the chain. In this chapter we look at what those benefits are and how different benchmarking schemes deliver them. We then go on to examine how the UK stacks up against other countries in terms of the benchmarking schemes available to farmers and agricultural professionals.

In an international study of agricultural benchmarking services carried out by Prospect Management Services for the Food Chain Centre (IGD) we noted that benchmarking services delivered some or all of the following benefits to farmers through a variety of features. The number in brackets represents our rating of the relative importance of the benefit, where 5 is high and 1 is low:

- (5) Low effort required in order to participate and benefit from benchmarking.
- (5) Financial improvements as a result of opportunities and weaknesses highlighted.
- (5) Demonstrates performance relative to other farm enterprises.
- (4) Credibility – of the system, its contents and thus the learning derived from the results.
- (3) Provision of a decision-making and planning tool which can be used when making strategic decisions.
- (3) Social benefits in terms of discussions with advisors, subjects of conversation with other farmers and agricultural professionals.
- (3) Social and business benefits as a result of participating in progressive farming approaches.
- (4) Immediacy of benchmarking reports – i.e. immediate answers to questions rather than waiting for weeks or months, by which time the impetus has been reduced.

- (2) Improved relations with lenders – agricultural banking specialists have welcomed benchmarking results as a means of assisting them in assessing loan risk.
- (4) Clarity of understanding one's own figures – by participating in the benchmarking process farmers develop a much deeper understanding of the finances of their business.
- (4) Increased sustainability – whether through recognising opportunities sooner or highlighting issues earlier.
- (4) Pinpointing weaknesses more clearly.
- (2) International comparisons to the UK, particularly important for those products where the market is global rather than local.
- (3) 24/7 availability of the system.

In addition to the benefits delivered to farmers, we saw that in some of the systems successful benchmarking services delivered benefits to other elements of the agricultural supply chain. The key benefits included an increased capability to assess financial risk, an increase in the levels of discussion about the benchmark findings with upstream and downstream supply chain partners, an overall improvement in the understanding of agricultural economics at a farm level and improved advisory services.

An increased capability to assess financial risk was most likely to be targeted at financial lending institutions, but was generally considered to be mutually beneficial to both farmers and lenders. Lenders would be provided with access to overall reports and would ask farmers for their individual data in order to compare the two and assess the lending risk.

Several schemes enabled farmers to share data directly with supply chain partners such as feed merchants; others spoke of farmers using the data in negotiations. In either case, benchmarking has been seen to add detail to the farmer's discussions with supply-chain partners. This improvement is expected to be connected to the impact that benchmarking is making upon the understanding of the realities and variances in economics at a farm level across government, advisory services and trade associations. This is enabling better informed strategy development and targeted policy initiatives. Furthermore, in those examples where benchmarking is delivered as a part of a broader agricultural advisory service, it can be seen to contribute to the development of packaged and staged support services which have the overall effect of improving farmers' fluency with farm economics (with the goal of addressing the current imbalance between economic and agronomic understanding amongst farmers).

Analysing the Features Provided by Different Benchmarking Systems

In order to be able to compare the benchmarking systems studied, the importance of the above farmer benefits was rated and the ability of the various features to deliver to these benefits analysed. Table 3.1 shows how features deliver farmer benefit.

The features provided by each benchmarking service were then assessed in order to provide a score. Table 3.2 summarizes this.

Table 3.1 Farmer benefits from mass participation benchmarking systems

Farmer Benefits

Features of Service	Low effort 5	£ Gain 5	Relative performance 5	Credible 4	Decision/planning tool 3	Social 3	Progressive 3	Immediacy 4	Loan support 2	Know own result 4	Increased sustainability 4	Pinpoint weakness 4	UK vs global 2	Always on tap 3	Score
Financial as well as physical	0	1	0	1	1	0	1	0	1	1	1	1	0	0	29
Enterprise level	0	1	0	1	1	0	1	0	1	1	0	1	0	0	25
Ease of use	1	0	0	1	1	0	0	1	0	1	0	1	0	1	27
Part of something else	1	0	1	1	1	0	1	0	0	1	0	0	0	0	24
Advisor involvement	1	1	1	1	1	1	1	0	0	1	1	1	0	0	40
Influencer	0	1	0	1	0	0	1	0	0	1	0	0	0	0	16
Leadership	0	1	0	1	0	0	1	0	0	1	1	0	0	0	20
Commercial provider	1	1	1	1	1	0	0	1	0	0	0	1	0	0	30
Group focus	0	1	1	1	1	1	1	0	0	1	0	1	0	0	31
Add-ons developed	0	1	0	1	1	0	1	0	0	1	0	0	0	0	23
User controls access	0	1	0	1	1	1	1	0	0	0	1	0	0	0	22
Drill down	0	1	1	1	1	0	1	1	0	0	0	1	0	1	35
Define own comparison group	0	1	1	1	1	0	1	1	0	1	0	1	0	1	39
Scenario data	0	0	0	1	1	0	1	1	0	0	1	0	0	1	21
Web based	1	0	1	1	1	0	0	1	0	0	0	1	0	1	31
International data	0	0	0	1	0	0	1	0	0	0	1	1	1	0	17

Table 3.2 Key features of mass participation benchmarking systems

Features of Service	Score	TEAGASC	Agrovision	Greenmount	Agrimetrics	Agrosoft	Finbin	Bank West	PigChamp	OABS	LEI
					Benchmarking Service						
Advisor involvement	40	1	0	1	1	0	1	1	0	1	1
Define own comparison group	39	0	1	0	0	1	1	1	0	0	0
Drill down	35	0	1	0	0	1	1	0	1	0	1
Group focus	31	1	0	1	0	0	0	1	0	0	0
Web based	31	1	1	1	0	0	1	0	1	0	0
Commercial provider	30	0	1	0	1	1	0	1	1	1	1
Financial as well as physical	29	1	1	1	1	1	1	1	0	1	0
Ease of use	27	1	1	1	1	1	1	1	1	1	1
Enterprise level	25	1	1	1	1	1	1	1	1	1	1
Part of something else	24	1	1	1	0	1	1	1	1	1	1
Add-ons developed	23	0	0	0	1	0	1	0	0	1	0
User controls access	22	0	1	0	0	0	1	0	0	1	0
Scenario data	21	0	0	0	0	0	0	0	0	0	0
Leadership	20	1	0	1	1	1	1	1	0	1	0
International data	17	0	1	0	0	1	0	0	1	0	0
Influencer	16	0	1	0	0	0	1	1	0	1	0
Total Score	**430**	**250**	**295**	**227**	**194**	**277**	**331**	**281**	**189**	**256**	**211**
% Score		58	69	53	45	64	77	65	44	60	49

Having devised a method for comparing the various agricultural benchmarking services around the world, the question we are faced with is how do we achieve mass participation?

WHAT REALLY MAKES A DIFFERENCE?

There were several elements of provision which were consistently considered to be key to achieving mass participation by farmers.

The provision of an advisor network was deemed to be the single most important factor in achieving high levels of high-quality participation. The ability to encourage and deliver benchmarking services within these relationships with farmers is seen to be a highly successful method in terms of raising awareness as to the benefits, assisting farmers in collating the necessary data, and interpreting the results. Furthermore, the feedback that advisors received during these conversations has been seen to lead to improvements in the services provided and enables them to get a handle on the challenges faced by farmers in the current economic climate.

The goal of achieving financial improvements in their farming business was often assumed rather than stated. However, the idea that the service is provided for farmers for their use in decision planning (rather than data gathered for exploitation by other elements of the supply chain) was explicit in the objectives of the organizations delivering benchmarking services, even if it was implicit in the benchmarking services themselves.

In the course of the research, we identified a series of related activities across different points in the supply chain, which could be seen to be driving the participation rates as well as increasing the value of the systems overall. We would group these factors into the following clusters:

* valued by other parts of the supply chain;
* graduating into greater complexity;
* connected technologies.

VALUED BY OTHER PARTS OF THE SUPPLY CHAIN

If a benchmarking scheme is valued by many parts of the supply chain then the pressure to participate comes from several directions at once. That is to say, if your competitors, customers and suppliers all use the system then it becomes a basic requirement for doing business in the sector and the key metrics become a regular component of decision making and negotiations.

Bank West has a commercial need for information as part of its risk assessment and management activities, but the findings are widely used by other farmers and banks. Finbin's service has always been fully publicly accessible. This has enabled influencers such as agricultural banks to access the data and develop their own uses for it. FinPack for Lenders is the net result. It would appear that both farmers and lenders benefit from this, as the quality of discussion around loans and expectations has been raised. It would be interesting to explore the potential in this area in the UK. This demonstrates the role that influencers such as the agri-business division of UK banks could play in encouraging participation in benchmarking schemes in the UK. Additionally, when the services are endorsed by trade organizations, as can be seen both in the UK and the US, the services can be marketed and promoted via the trade organization's own programmes, thus increasing farmer awareness and participation. In the UK, producer groups (related to the trade association) are a key point of delivery of services for the pig sector and membership is mandatory.

Collaboration across the supply chain and between trade bodies has been seen to be a means of fast-tracking participation in South Africa. The deregulation of the deciduous fruit industry in South Africa in 1997 and the ensuing absence of benchmarking information made it clear to the sector that a structured approach to data collection was needed. The leadership of the Deciduous Fruit Producers Trust combined with the independence of the OABS (Optimal Agricultural Business Systems) provided a strong and trusted mechanism for data collection and analysis.

Delivery of the service by a non-competing and independent organization enables the scheme to act as a trusted intermediary between competing commercial firms. This enables all parties to contribute far more detailed and sensitive information than they would share directly, as well as providing a route for questions and dialogue on related issues.

Advocacy and support is a critical element in achieving mass participation. In the USA, it is interesting to note that where Finbin is launched in states without an advisor base the uptake is much slower. This would seem to confirm the view that advisor assistance is key. The French approach is to support benchmarking via associated consultants; this increases CER France Allier's capacity to cover both geography and sectors. Advocacy also takes the form of personal leadership by the senior farmers and entrepreneurs in a sector. These individuals combine use of the system internally with the pursuit of a long-term vision for the sector as a whole.

The long-term delivery of information *for* the sector delivered *by* the sector can be seen to be a factor in the development of credibility. In Ireland, Profit Monitor has now been running for a sufficiently long time that the processes have become well integrated into farmer practice and the annual diary of work to be undertaken. By contrast, LEI collect data which the Dutch government requires for EU purposes; benchmarking is an add-on designed to deliver value back to the participating farmers (*a quid pro quo*). This prioritizes data collection for the government not the delivery of value to farmers. Those systems which are by the sector, for the sector are less likely to suffer from this issue of trust and lack of flexibility from which LEI suffers.

GRADUATING INTO GREATER COMPLEXITY

The essence here is that farmers need a way in to benchmarking, and that way in needs to be relevant to the challenges they currently face whilst allowing farmers to become involved in greater depth as they integrate benchmarking into their decision-making processes.

This means that in order to achieve mass participation successful schemes employ a series of routes of engagement, starting with broad overviews and following through to depth in specific areas. As many farmers need to improve their financial management systems both in order to contribute data into benchmarking and obtain comparisons from it, this often runs in parallel with the routes into benchmarking. A sensitivity surrounding disclosure of financials is to be found amongst UK farmers. Developing a smaller set of less sensitive ratios which can provide an entry point into such discussions becomes an important tool in establishing conversations around economy as well as agronomy. Improved financial management systems support these new conversations.

A commercial entity has a financial driver for continuing to extend the services it can provide to its client base and Agrimetrics is no exception. The company provides a range of services, starting with a review of a client's cost base but potentially expanding into analyses of plant and sales. This provides a route in to use of benchmarking data as well as opportunities to work with firms to help them achieve greatest value from the services they procure from Agrimetrics.

An example of the quantum leap that can be made can be seen in the pig industry. Pig farmers are already highly numerate and familiar with the details of their business. This has made uptake and use of the comparison tools much easier. The feedback loop between data generated and management information used in decision making has become much faster – a necessity in any industry faced with high levels of competition and price sensitivity.

Once the systems and processes are in place to enable farmers (and other members of the supply chain) both to contribute to, and to use, the management information that benchmarking provides, these systems then go on to consider connections to other management tools. Agrovision is one of a number of examples where benchmarking is piggy-backed on other farm management tools. This approach increases ease of use and reduces barriers to participation. In Ireland the benchmarking systems have been connected to accounting systems.

Whilst complexity enables users to drill down into the detail, ease of use continues to be a key driver of ongoing participation. This is being achieved in a number of different ways: advisor-delivered services were one route, benchmarking systems piggy-backed on farm management software systems was another common mechanism. The latter has the advantage of increasing the immediacy of access to the results, as benchmarking is commonly only one click away and does not require data re-entry.

The South African fruit industry is dependent upon demand from foreign markets. To achieve stable and high-quality collaboration between industry players requires them to consider the competition as being other export-focussed fruit producing countries rather than the guy next door. Whilst co-opetition is well understood in business generally, it is still relatively new to the agricultural sector. South Africa provides a good example in this regard.

CONNECTED TECHNOLOGIES

As benchmarking systems require a combination of user participation and large databases it makes a lot of sense to provide these services via web applications rather than software applications. In moving from software applications to web-based applications a variety of other connections of technology are made possible.

One key participation driver is the ease with which data is captured into the benchmarking system. Sourcing data from farm accounts saves farmers entering data separately into the benchmarking forms. Whilst UK farmers do not subscribe to the types of system available in France they do often use farm accountants. There would be value in exploring whether improved connections to these influencers could raise benchmarking participation levels in the UK. More recent work with farm accountancy software providers has demonstrated the benefits of connecting these tools with Profit Monitor.

This work will continue to develop. It is expected that this will lead to an electronic connection between the systems. Elsewhere the piggy-backing of the comparison element on a successful piece of management software has been key to success. For example, the provision of benchmarking services within a broader advice and education structure enables Teagasc and its advisors to signpost farmers to the benchmarking system.

Among the systems studied, web-delivered benchmarking services are in the majority and are considered to be the way forward both in terms of data collection and reporting. In particular, the advantage of 24/7 access and immediate answers to questions asked of the system by users was deemed to be very important. Web delivery also enhances the provision of comparisons and the detail of drill-down information accessible.

The above elements all benefit from the support of a credible provider with a commitment to the progression of farming practice and increased sustainability of farming business within a local, regional, national and international context.

Conclusion

In the course of studying some 15 benchmarking systems around the world we noted not only the features and benefits which were key drivers towards mass participation, but also a series of trends which suggest where the future developments may lie in this area of work.

The trend towards integrated use of benchmark data throughout the supply chain is a strong indication of the wide benefits that benchmarking can deliver. In order to make the most of this opportunity, benchmarking systems are providing a greater variety of entry routes and the opportunity to connect up and correlate information provided by different parties within the same supply chain.

A part of the process of increasing the accessibility of benchmarking is not only to open it up to other parts of the supply chain, but also to make it available 24/7 via the web – both for data entry and results viewing purposes.

Furthermore, once benchmarking becomes a web application rather than a stand-alone software application then the options for connecting it to other data sources such as farm accounts data arise. This feature can be seen to appeal to a whole tranche of farmers who have not previously participated in benchmarking but would be much more likely to do so when it is integrated into their other activities.

Our anecdotal evidence suggests that benchmarking around the world covers some 5–15 per cent of farmers. Based on the above trends we would expect this number to increase steadily over the next decade as participation becomes easier and the supply chain members start to demand it of one another.

4 DairyBase: Building a Best Practice Benchmarking System

NICOLA SHADBOLT
Massey University

In 2003 a voluntary-based industry group in New Zealand, calling itself the KPI Working Group, formed to discuss and address the fragmented approach to measurement of business performance that existed in the dairy industry. It was recognized that not only was the data fragmented and not always robust but that there were also inconsistencies in both the terminology and calculation of key performance indicators (KPIs). The objective the group set itself was to develop a co-ordinated approach to provide sound, robust data and consistent benchmark calculations which would provide increased clarity of data for the dairy industry and benchmarks that could be relied upon.

The voluntary working group consisted of representatives from the NZ Institute of Chartered Accountants, Dexcel, NZ Institute of Primary Industry Management, Massey University, Fonterra and trading banks.

Background

It is of interest first to note how industry standards have developed in New Zealand and what role the various organizations have played in this development.

The NZ Institute of Chartered Accountants (formerly New Zealand Society of Accountants) has always played an active role in farm management accounting. McEwen (1965) documents the process by which an Agricultural Development Conference resulted in the following recommendations:

1. That the NZ Society of Accountants (NZSA) convene a committee to revise the form of accounts and code of terminology in the 1961 Research Report of Farm Accounting to provide forms for use by the farmer to record essential management and financial information during the year.
2. That to ensure the widest possible adoption of the recommendations regarding minimum standards for farm accounting a publicity campaign among farmers and accountants be sponsored by the NZ Society of Accountants, the government

producer boards, Federated Farmers and others, including lending institutions and farm improvement clubs.

At that time New Zealand's 73,000 farmers earned over 90 per cent of the country's total overseas earnings and it was noted with concern that there was a serious lack of information on the economic aspects of farming.

The result of implementing the above recommendations was the publication of NZSA's *Farm Accounting in New Zealand* (commonly referred to as 'The Green Book') in 1968, in which an agreed chart of accounts was presented, as were recommended formats for various accounting reports including a cash flow statement. It is of interest to note that this publication outlines an operating profit, or Economic Farm Surplus statement, as the method for providing comparisons between one farm and another and between different years on the same farm. The publication recommends three major reporting statements as critical to business analysis:

- the farm working account (now known as the Statement of Financial Performance);
- the Cash Flow Statement;
- the Economic Farm Surplus.

In the preface to this NZSA (1968) publication it is stated '...no longer is it sufficient for the accountant to produce only historical records and taxation returns – he must be looking ahead and fulfilling his role as his client's financial adviser'. It also notes how the changeover to decimal currency in 1967 and the increasing use of computers 'presages a climate of change and progress and the need for more precise planning of farming operations'.

In 1977 the NZSA produced a subsequent publication, *Management Accounting for the New Zealand Farmer* (NZSA, 1977). In this it was stated that accounts prepared on a purely historical cost basis are misleading to the user and that there was an increased emphasis on accounting to provide information for business management that was essential to sound decision making. The Society recommended a move away from accounts drawn up largely for tax assessment, the adoption of net current values for assets and the abandonment of tax values for livestock. A subsequent NZSA publication, *Financial Reporting for Primary Producers*, was produced in 1989 to update members on the continuing changes in financial reporting requirements (Clarke, 1989). Its purpose was to recommend accepted accounting principles for primary producers 'with a view to providing guidance on financial reporting and valuation policies and techniques for primary producers and their financial advisers'. Again a sample set of statements is presented, including cash flows, but no chart of accounts is included this time and, as with the 1977 publication, no mention is made of Economic Farm Surplus. It presents financial reporting as being primarily historical but suggests a sound accounting and financial reporting system provides a greater degree of precision that will enable better assessment of unprofitable areas and areas where economies can be made. Subsidies were removed from NZ agriculture in 1984 so it is not surprising that the publication presents producers' ability to make sound financial decisions as becoming increasingly more important as they deal with variable input costs and volatile market conditions, debt levels and interest rates.

A consistent theme throughout these publications has been the recommendation that accountants produce a cash flow statement in conjunction with other financial statements, but this has never become a legislated requirement. McEwen (1965) identified the cash flow statement as a restatement of the accounts in the form of total sales and expenses ignoring the profit concept of accounting; he believed it was in the cash flow form that his farmers thought about finances. He also pointed out how the farm budgets used are simply a projection of the cash flow statement for the following year, so providing a cash flow statement of the year that has been assists in the farmer's projection of the year to come. Clarke (1989) defines the task of the cash flow statement as being to provide information about the operating, financing and investing activities of an entity and the effects of those activities on cash resources.

However, despite this early work and subsequent recommendations by the NZSA, Angus (1991) identified that the conventional presentation of accounts was still failing to communicate clearly a meaningful cash result. Angus (1991) states that while most farming clients are well served by their accountants in the area of legitimately minimizing tax, the 'simple objective of defining if earnings exceed spending has been lost sight of'.

However, since 1965 a dedicated group of farm accountants has developed in New Zealand; this group has put into practice many of the recommendations of the various NZSA publications and many of them have also developed various forms of benchmarking for their clients, analysing the cash result, the profitability and the equity change of their clients and comparing each result with group averages.

In parallel with these developments in the accounting profession, and perhaps because of them, other rural professionals (including the Ministry of Agriculture in the 1970s) have also developed various methods of financial reporting. Bankers tend to focus very closely on the cash position of their clients; they link this to changes in stock numbers and capital purchases to determine if their client's risk status has changed. They also monitor asset values to determine client debt to asset ratios and, inversely, the increasing or reducing risk of their lending portfolio. Farm consultants commonly assist farmers with their cash budgets so also require details on the cash position of previous years. In the absence of meaningful cash flow statements both they and farm financiers must complete accounts analyses (cash reconciliations) to determine historical cash results on which to base or compare projections. As farm consultants are also often involved in benchmarking for a group of clients they have tended to calculate economic farm surplus (various versions based on the NZSA [1968] recommendation) and other efficiency ratios (return on assets, return on equity and various per hectare, per stock unit and per kg output measures).

Over time, the definitions of such measures altered at the whim of the people involved and the connection with a common standard or definition was lost. Their varying academic backgrounds (accountancy, farm management or valuation) largely determined the emphasis they placed on liquidity, profitability, efficiency, taxation and equity, and the reliability and accuracy of each calculation.

Operating profit, often termed Economic Farm Surplus in New Zealand, is calculated both for dairy farms and for sheep and beef cattle farms in annual statistics collected by the respective industries (The Economic Service, 2006).

The Process

Despite the wide range of measures and definitions used by the various members of the group and a high level of 'patch protection', the working group made good progress in the first 12 months, deciding on KPIs and their standardization. Most members of the group provided a type of benchmarking service to their clients, in which considerable investment had been made in data collection, analysis and interpretation. However, all members saw the benefit in pooling their skills and the farm data to enable a national service to be developed. In October 2004 funding for the project was granted by Dairy Insight (the dairy farm levy investment manager at that time). This allowed the working group to proceed with the development of the software, the web interface, the reports and database systems, and procedures to establish DairyBase.

The buy-in and contribution from all members of the group has been the key reason for the project's success to date. Ultimately the project will only be successful if rural professionals use the database and adopt the calculations and terminology as the industry standard. It is critical that the benchmarks are produced from a system which has integrity and will allow meaningful comparisons. The group determined that integrity resulted from having trained individuals entering standardized and verified data that meets specified quality standards. The volume of data, or number of data sets entered from different farms, must be high enough to ensure an accurate representative sample. By 2008 there were over 1,000 farm businesses analysed by DairyBase and the target is to have 5,000 (50 per cent of farm businesses) analysed by 2010.

The key objectives of DairyBase are to:

* standardize terminology, calculations and reporting of key KPIs;
* provide sufficient volumes of reliable data for farm comparisons;
* develop a national database for the dairy industry that will provide robust national and regional data for different farm types. This includes producing an annual publication of industry trends;
* provide improved aggregate data to measure industry progress and for R&D purposes.

Accredited rural professionals enter farm physical and financial data. It is anticipated that accountants will enter most of the data as they finalize each year's Annual Financial Statements. If accountants do not enter the data it can be entered by accredited consultants or bankers.

Rural professionals are not permitted to enter data without authorization from the farm business owner. The farm business owner is able to authorize any one or more rural professionals to enter data into the system. The initial data is entered over the internet to a validation or scratchpad area. Once the data has been validated or passed through a series of checks it is transferred into the actual database. Figure 4.1 summarizes the DairyBase process.

Reports are generated after data has been validated and committed to the database. The reports produce data for the individual farm business and the data for a chosen benchmark group. Examples of the available reports are given at the end of this chapter.

Reports of aggregate (not individual farm business) data will be made available to industry bodies as requested. Market research carried out at the commencement of this project confirmed that a National Database for the dairy industry to provide information to industry for R&D and planning purposes, and also provide a basis for benchmarking, was supported by the majority.

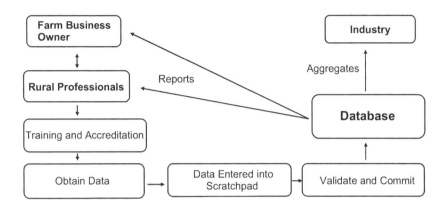

Figure 4.1   The DairyBase process

LEVEL ONE PHYSICAL AND FINANCIAL REPORTS

These reports focus on a physical summary then KPIs in the three critical areas:

1. Cash (liquidity);
2. Profit;
3. Wealth creation.

The emphasis on cash noted by McEwen in 1965 is still as valid today for many farmers and is an essential financial management skill at the operational level (Shadbolt and Gardner, 2006). The focus on profit and efficiency includes the operating profit, return on assets and return on equity as well as the key Du Pont drivers of operating profit margin and asset turnover. Results are stated also on a per hectare, per cow and per kg milksolids basis. Delivery to these measures is the result of good financial management at the tactical level as managers make revenue generation and cost control decisions as the season unfolds. Wealth creation is recognized as a key financial outcome at the strategic level for many farm businesses and is reliant on a realistic estimate of asset values at opening and closing (Shadbolt and Rawlings, 2001). The important distinction is also made between wealth created from profit retained and invested in the business and that achieved as a result of changing asset (land and shares) values. Various solvency and debt servicing capacity measures are also included, to ensure the vulnerability of the business is understood.

More in depth Level Two physical data can also be collected to provide more physical analysis of the farming system.

Physical Data Summary

_Farm Ltd	Date Printed: 4 August 2008
Dairy Season ended: 2007	Farm ID:

Dairy Co Supplied:	Fonterra		
Production System:	3		
Business Type:	Owner Operator	Balance Month:	May
Calving Season:	Spring Only	Milking Interval:	Twice a day
Winter Milk:	No	Organic:	No
Region:	Marlborough – Canterbury	District:	Selwyn
NIWA 10 Yr Av Rainfall (mm):	720	Season's rainfall (mm):	650
% Milking Area Irrigated:	More than 30%		
Farm Dairy Type:	R50	Predominant Soil Type	

Stock

Predominant dairy breed:	Friesian
Peak Cows Milked:	1,020
Stocking Rate (Cows/ha):	3.7
Replacement Calves Reared:	215
Non-replacement Calves Reared:	0

Land Area (ha)

Total Dairying area:	280.0
Less Ungrazeable are:	6.0
Effective Dairying area:	274.0
Less Defined Young Stock area:	0
Milking area:	274.0
Dairy Run – off effective area:	0.0
Non – dairy effective area:	0.0

Labour

Full time paid labour equivalents:	7.2
Full time unpaid labour equivalents:	0.0
FTE unpaid management:	0.0
Total FTE's:	7.2
Milking Cups per FTE:	7

Production	Total	Per ha	Per cow	Composition
Milk Litres:	5,286,482	19,294	5,183	
Fat kg:	257,463	940	252	4.9%
Protein kg:	202,555	739	199	3.8%
Milksolids kg:	460,018	1,679	451	8.7%
Non-replacement calf milk (l):	0			
Non-replacement calf MS (kg):	0			

Number in Benchmark Group:	60
Benchmark Group Selected by:	Region: Marlborough-Canterbury
	Farm business type: 1 Owner Operator
Benchmark Group Ranked by:	

Data entered by:	Financial:	Extended Physical:

Figure 4.2   Sample DairyBase farmer report 1

	Key Performance Indicators	
_Farm Ltd		Date Printed: 4 August 2008
Dairy Season ended: 2007		Farm ID:

Number in Benchmark Group:	60

Benchmark Group Selected by:	Region: Marlborough – Canterbury
	Farm business type: 1 Owner Operator

Benchmark Group Ranked by:

Farm Physical KPI's	2006–07		2005–06		2004–05	
	Farm	Benchmark	Farm	Benchmark	Farm	Benchmark
Cows/ha	3.7	3.5	3.7	3.4		
Kg Milksolids/ha	1,679	1,422	1,507	1,350		
Kg Milksolids/cow	451	403	405	398		
Cows/FTE	143	154	146	145		
Kg MS/FTE	64.338	62.049	59.113	57.815		

Profitability	2006–07		2005–06		2004–05	
Dairy	Farm	Benchmark	Farm	Benchmark	Farm	Benchmark
Gross Farm Revenue/ha	7,284	6,579	6,139	6,138		
Operating Expenses/ha	4,875	5,256	4,486	4,467		
Operating Profit (EFS)/ha	2,409	1,341	1,671	1,464		
Gross Farm Revenue/kg MS	4.34	4.64	4.07	4.55		
Operating Expenses/kg MS	2.90	3.70	2.96	3.46		
Operating Profit (EFS)/kg MS	1.44	0.94	1.11	1.08		
FWE/kg MS	2.65	3.15	2.56	2.75		
Operating Profit Margin %	33.1%	20.3%	27.2%	23.8%		
Asset Turnover %	17.9%	15.7%	12.9%	14.3%		
Operating Return on Dairy Assets %	5.9%	3.3%	3.5%	3.5%		

LIQUIDITY	2006–07	2005–06	2004–05
Net Cash Income	1,965,683	1,589,313	
Farm Working Expenses	1,221,047	1,057,672	
Cash Operating Surplus	744,636	531,641	
Discretionary Cash	306,999	143,392	
Cash Surplus/Deficit	-49,869	-122,660	

TOTAL WEALTH	2006–07	2005–06	2004–05
Closing Dairy Assets $	11,946,802	11,162,406	13,077,388
Closing Total Assets $	11,946,802	11,162,406	13,077,388
Closing Total Liabilities $	4,875,608	5,000,739	5,301,383
Closing Total Equity	7,071,194	6,161,667	7,775,550
Growth in Equity $	909,527	967,358	
Growth from Profit	107,451	-69,519	
Growth from Capital	802,076	1,036,877	

	2006-07		2005-06		2004-05	
	Farm	Benchmark	Farm	Benchmark	Farm	Benchmark
Growth in Equity %	14.8%	5.9%	12.4%	4.1%		
Debt to Assets %	40.8%	46.0%	38.4%	38.6%		
Opening Liabilities/kg MS	10.87	13.41	12.84	12.10		
Closing Liabilities/kg MS	10.60	14.85	13.17	13.35		

Comment: Assets include Land and Building values calculated using revalued capital values.

Figure 4.3   Sample DairyBase farmer report 2

Cash Flow and Profitability		
_Farm Ltd		Date Printed: 4 August 2008
Dairy Season ended: 2007		Farm ID:

Number in Benchmark Group:		60
Benchmark Group Selected by:		Region: Marlborough – Canterbury Farm business type: 1 Owner Operator
Benchmark Group Ranked by:		

CASH	$/KG MS	$	NON CASH ADJUSTMENTS	$	CASH + NON CASH	$
DAIRY SALES						
Net Milk	4.09	1,879,377			Net Milk	1,879,377
Net Livestock	0.14	66,969	+ Value of Change in Dairy Livestock	20.157	Net Livestock	96,883
Other Dairy	0.04	19,610			Other Dairy	19,610
NET CASH INCOME	4.27	1,965,683			**DAIRY GFR**	1,995,870

CASH FWE	$/KG MS	$	NON CASH ADJUSTMENTS	$	OPERATING EXPENSES	$
Wages	0.51	235,879	+Labour Adj	0	Labour Expenses	235,879
Stock Expenses	0.23	106,130			Stock Expenses	106,130
Supplementary Feed	0.40	185,870	+Feed Inventory Adj	0	Total Supplement Expenses	185,870
Grazing and Run – off	0.54	249,934	+Owned Run – off Adj	0	Total Grazing and Run-off	249,934
Other Working Expenses	0.81	373,706			Other Working Expenses	373,706
Overheads	0.15	69,528	+Depreciation	114,644	Total Overheads	184,172
FARM WORKING EXPENSES	2.65	1,221,047			**OPERATING EXPENSES**	1,335,691

CASH OPERATING SURPLUS	1.62	744,636	NET ADJUSTMENTS	-84.457	**DAIRY OPERATING PROFIT (EFS)**	$660,179

TOTAL BUSINESS				TOTAL BUSINESS	
-Rent	0.00	0		-Labour Adjustment	0
-Interest	0.83	382,191		-Owned Run off Adjustment	0
-Tax	0.12	56,820		-Non-Dairy Operating Profit	0
-Net non Dairy Cash Income	0.00	0		-Net Off Farm Income	1,374
-Income Equalisation	0.00	0		-Extraordinary Expenses	0
-Net off-farm income	0.00	1,374		-Rent (excel run-off)	0
DISCRETIONARY CASH	0.67	306,999		-Interest	382,191
				Business Profit Before Tax	279,362

Applied to:					
-Net Capital Transactions	0.15	66.777			
-Net Debt	0.38	175,000		-Drawinga	115,091
-Net Drawings	0.25	115,091		-Tax	56,820
-Extraordinary Expenses	0.00	0		**EQUITY GROWTH FROM PROFIT**	107,451
-Net Funds Introduced	0.00	0			
CASH SURPLUS/DEFICIT	-0.11	-49,869			

Comments:

Figure 4.4   Sample DairyBase farmer report 3

	Financial Detail
_Farm Ltd	Date Printed: 4 August 2008
Dairy Season ended: 2007	Farm ID:

Number in Benchmark Group:	60
Benchmark Group Selected by:	Region: Marlborough – Canterbury Farm business type: 1 Owner Operator
Benchmark Group Ranked by:	

	Total $		$ Per Kg MS		$ Per Ha		$ Per Cow	
GROSS FARM REVENUE (GFR)	Farm	% of GFR	Farm	Benchmark	Farm	Benchmark	Farm	Benchmark
Net Milk Sales	1,879,377	94.2%	4.09	4.25	6,859	6,046	1,843	1,715
Net Dairy Livestock Sales	66,969	3.3%	0.14	0.28	243	402	65	114
Value if Change in Dairy Livestock	30,157	1.5%	0.07	0.08	110	119	30	34
Other Dairy Revenue	19,610	1.0%	0.04	0.02	72	30	19	5
Dairy Gross Farm Revenue	1,995,570	100.0%	4.34	4.64	7.284	6,579	1,957	1,871
Non-Dairy Cash Income	0	0.0%	0.00	0.01	0	18	0	5
Value of Change in Dairy Livestock	0	0.0%	0.00	0.00	0	2	0	1
TOTAL GROSS FARM REVENUE	1,995,570	100.0%	4.34	4.65	7.284	6,616	1,957	1,877

OPERATING EXPENSES								
Labour Expenses								
Wages	235,879	11.8%	0.51	0.58	861	826	231	234
Labour Adjustment – Unpaid	0	0.0%	0.00	0.04	0	51	0	14
Labour Adjustment – Management	0	0.0%	0.00	0.11	0	155	0	44
Total Labour Expenses	235,679	11.8%	0.51	0.73	861	1,032	231	293
Stock Expenses								
Animal Health	47,641	2.4%	1.10	0.15	174	219	47	62
Breeding and Herd Improvement	27,551	1.4%	0.06	0.09	101	126	27	36
Farm Dairy	15,554	0.8%	0.03	0.04	57	61	15	17
Electricity (Farm Dairy, Water Supply)	15,394	0.8%	0.03	0.09	56	122	15	35
Total Stock Expenses	106,130	5.3%	0.23	0.37	387	528	104	150
Feed Expenses								
Supplement Expenses								
Net Made,Purchased,Cropped	184,427	9.2%	0.40	0.62	673	880	181	250
Less Feed Inventory Adjustments	0	0.0%	0.00	0.02	0	33	0	9
Calf Feed	1,443	0.1%	0.00	0.01	5	21	1	6
Total Supplement Expenses	185,870	9.3%	0.40	0.65	678	868	182	246
Grazing and Run-off Expenses								
Young and Dry Stock Grazing	249,934	12.5%	0.54	0.33	912	469	245	133
Winter Cow Grazing	0	0.0%	0.00	0.03	0	42	0	12
Run-off Lease	0	0.0%	0.00	0.07	0	98	0	28
Owned Run – off Adjustment	0	0.0%	0.00	0.05	0	65	0	18
Total Grazing & Run-Off Expenses	249,934	12.5%	0.54	0.48	912	674	245	191
Total Feed Expenses	435,804	21.8%	0.94	1.13	1,591	1,542	427	437
Other Working Expenses								
Fertiliser	132,586	6.6%	0.29	0.28	484	403	130	114
Nitrogen	23,659	1.2%	0.05	0.10	86	147	23	42
Irrigation	148,157	7.4%	0.32	0.14	541	202	145	57
Regrassing	9,363	0.5%	0.02	0.05	34	66	9	19
Weed and Pest	518	0.0%	0.00	0.02	2	29	1	8
Vehicles	7,857	0.4%	0.02	0.07	29	96	8	27
Fuel	12,022	0.6%	0.03	0.05	44	76	12	21
R&M –land and buildings	26,046	1.3%	0.06	0.15	95	217	26	61
R&M – plant and equipment	10,848	0.5%	0.02	0.08	40	112	11	32
Freight and General	2,620	0.1%	0.01	0.03	10	45	3	13
Total Other Working Expenses	373,706	18.7%	0.51	0.98	1,364	1,393	366	395
Overheads								
Administration	41,132	2.1%	0.09	0.07	150	99	40	28
Insurance	8,923	0.4%	0.02	0.03	33	38	9	11
ACC	8,739	0.4%	0.02	0.02	32	34	9	10
Rates	10,734	0.5%	0.02	0.03	39	47	11	13
Depreciation	114,644	5.7%	0.25	0.38	418	543	112	154
Total Overheads	184,172	9.2%	0.40	0.53	672	760	181	216
Total Dairy Operating Expenses	1,335,691	66.9%	2.90	3.70	4,875	5,258	1,310	1,491
Non- Dairy Operating Expenses	0							
TOTAL OPERATING EXPENSES	1,335,691	66.9%	2.90	3.70	4,875	5,262	1,310	1,493

OPERATING PROFIT								
Dairy Operating Profit (EFS)	660,179	33.1%	1.44	0.94	2,409	1,341	647	380
Non-Dairy Operating Profit	0	0	0.00	0.01	0	14	0	4
TOTAL OPERATING PROFIT	660,179	33.1%	1.44	0.95	2,409	1,354	647	384

Figure 4.5   Sample DairyBase farmer report 4

Summary

There is a need to provide farmers and wider industry players with information on liquidity, profitability and wealth creation/loss as it occurs on farms from year to year. No one measure is more important than another and each provides relevant information useful for both off- and on-farm decision-making. Critical areas that required consistency in how they were determined included the value of family labour and management, changes in feed inventory and the value of land and buildings. Indicators of success for both the property and the farming businesses was needed to ensure a holistic evaluation was made of overall investment strategy. DairyBase enables timely analyses carried out as close to the end of the financial year as possible; this in turn ensures the early creation of benchmarks that can then be used by farmers to compare their businesses with chosen benchmark groups.

DairyBase is an example of how inter-disciplinary groups can work towards a common goal and suggests a framework for farm analysis that could be used internationally.

References

Angus (1991), *The Cashflow Statement in Farm Accounts – Lets Make it Meaningful*. A project in fulfilment of the requirements of the NZ Society of Farm Management Study Award.

Clarke, M. (1989), *Financial Reporting for Primary Producers; for the Primary Sector Accounting Sub-Committee* (Wellington: The New Zealand Society of Accountants).

Dexcel (2006), *Economic Survey of New Zealand Dairy Farmers 2004–2005* (Hamilton: Dexcel Ltd).

FFSC (1997), *Financial Guidelines for Agricultural Producers*. Recommendations of the Farm Financial Standards Council. (Revised) December, 1997.

McEwen (1965), *Farm Management Accounting*. An address presented to the New Zealand Society of Accountants National Convention, Christchurch, NZ, 19 March, 1965.

NZSA (1968), *Farm Accounting in New Zealand*. Prepared by The Farm Research Committee of the New Zealand Society of Accountants' Board of Research and Publications (Wellington: The New Zealand Society of Accountants).

NZSA (1977), *Management Accounting for the New Zealand Farmer* (Wellington: The New Zealand Society of Accountants).

Shadbolt, N. and Gardner, J. (2006), Farm investments: alternative ownership structures that address the liquidity versus profitability conundrum, *Journal of International Farm Management* Vol. 3 No. 3 July 2006 <http://ifmaonline.org/pdf/journals/Vol3Ed3_Shadbolt_Gardner.pdf>.

Shadbolt, N.M. and Rawlings M. (2001), *Successful Benchmarking by Balanced Planning and Identifying Key Performance Indicators for Goal Attainment in Dairy Farming*. DRDC Australia Project MUNZ001.

The Economic Service (2006), *The New Zealand Sheep and Beef Farm Survey* (Wellington: The Economic Service of New Zealand).

Benchmarking for Better Practice in Farming

5 Benchmarking Clubs and Business Improvement Groups

Introduction

Benchmarking, as an activity distinct from the setting of standards for comparative analysis that has been well established in agriculture for over 100 years, and has been described as 'the search for industry best practices that will lead to superior performance' (Camp, 1989). Benchmarking, as understood as a business term, has evolved since the late 1980s in manufacturing industries through what is termed 'reverse engineering' through to competitive benchmarking, process benchmarking, strategic and then global benchmarking. Against this typology, benchmarking in agriculture is typically still in the traditional reverse engineering and cost reduction stage: forming a diagnosis of problems and then going back to change the operations of the business. Examples of this might be a high level of machinery cost leading to the decision to move to low- or non-tillage methods of cultivation for arable crops, or to increase throughput in a dairy by altering layouts and scheduling of milking sheds.

However, the use of benchmarking in agriculture is distinctive. Firstly, it is a collaborative activity centred on groups and has a social dimension. Secondly, benchmarking in this fashion has led to significant changes in operational practice and the identification of new revenues as well as cost savings, as demonstrated in the case studies referred to in this chapter. Previously unsustainable businesses have become economically sustainable. Because the benchmarking takes place against a small number of (usually) geographically related businesses, 'best practice' is not an accurate term for this type of benchmarking, and so the relevant section has instead been entitled Benchmarking for Better Practice.

The other distinctive feature of small group benchmarking in agriculture is that it drives learning at a number of different levels. Those not conversant with their own accounts have become cost aware and gained understanding of the interactions between operations and profits. In groups which do not share financial information the awareness may come through seeing how another farmer attracts custom to a farm shop, for example, or manages machinery more effectively. Kyrö (2003) states that a newcomer on the evolutionary path towards sophistication in benchmarking is 'competence' or 'learning benchmarking'. The term 'bench learning' has also been applied to this approach.

The philosophy behind this, as described by Kyrö (2003), is that 'the foundation of organisational change processes lies in the change of actions and behaviour of individuals and teams'. Learning *from* others, she goes on to say, led to the idea of learning *with* others through networks of organisations, leading to a new form of benchmarking – network benchmarking. So whilst the type of benchmarking being carried out in farm discussion or benchmarking groups could be described as benchmarking for better practice, the concept of benchmarking as networks and loci of learning activities is itself seen as best practice in benchmarking.

These approaches of learning and networking as central to benchmarking are exactly what has been happening in agriculture for the last 40 or more years, and for longer still if the farmers groups in nineteenth-century America or model farms in Europe which date back even further are considered in this light. Agriculture is therefore at a level of collaborative sophistication in benchmarking not yet reached by many other industries. It is both adaptive and generative in terms of driving innovation. This is where the potential lies for sustainable change through benchmarking in farming.

Farmers Groups and the Search for Better Practice

In the US, the development of farm bureaus was, as shown in Chapter 2, a means through which comparative analysis spread in the twentieth century. There was another development outside the US, though, where smaller groups of farmers got together to compare their ideas and data with each other. These discussion groups – sometimes known as business improvement groups or benchmarking clubs – have been the result in some cases of government or levy board sponsorship, in other cases of consultants offering additional services, and in some cases have been driven by individual farmers. The concept of discussion groups has been established in the UK since around 2000, and the idea has crept into the US, but the practice has been around for far longer in countries such as New Zealand and Australia, and across Europe. A Nuffield Farm Scholar Study in 1996 (Edwards, 1996) found well-established discussion groups in Denmark, the Netherlands and France, as well as New Zealand and Australia. Furthermore, the use of model or monitor farms is found in New Zealand and other areas, notably the Monitor Farms Programme in Scotland which was established in 2004 to replicate the New Zealand experience in Scotland (Fettes, 2006).

New Zealand Farm Improvement Groups and Monitor Farms

New Zealand is usually held up as the model for farm discussion groups. The discussion group concept appears to have been developed by advisors from Lincoln University on the South Island in the 1950s. The following extracts from the 1966 Encyclopaedia of New Zealand show the history:

Dairy Production and Marketing Board

This Board employs 13 consulting officers, two in Northland, four in South Auckland, one in the Bay of Plenty, two in Taranaki, two in Wellington – Hawke's Bay, and two in the South Island. These officers give an advisory service to dairy farmers like that given by the Farm Advisory Division of the Department of Agriculture. They cover such matters as calf rearing, bull selection, pasture management, milking techniques, etc. Being few in number, they prefer the group method of giving farm advice, organising many field days, and pioneering in New

Zealand the technique of the farm-discussion group. The latter were begun in 1952 and comprise groups of six to nine farmers who are prepared to meet periodically to discuss their farming problems and the application of research findings to farming practice. These discussion groups meet regularly in all areas covered by the consulting officers. Meetings take the form of round-table discussions with the subject decided beforehand, being usually one of topical interest. Sometimes outside speakers (usually specialists) address the group and lead the discussion on their work and its application to farming. Groups also hold field days on members' farms and sometimes visit research stations for on-the-spot talks with research workers.

Farm Improvement Clubs

In 1949 the Department of Agriculture was approached by members of Federated Farmers in the Waikato district to ask if an advisory officer could be appointed to act as adviser to 40 to 50 farmers, to consult with them, and draw up plans for improving production from their farms. Adequate management and accounting records would be kept so that the economic value of the improvement programmes could be assessed. The Department was unable to agree to the appointment of a special adviser for this work, but a modified scheme was agreed to whereby an advisory officer acted as adviser to a small group of five farms and an advisory officer in economics kept the necessary accounting and management records. As a result of this, three small groups of five farms each were started in the Waikato in 1950.

Some farmers, however, still considered that there should be some scheme whereby 40 to 50 farmers could be served by an adviser at the farmers' own expense. The first move was made in 1952 with the incorporation of the Franklin Farm Improvement Club. This club now employs two officers, each one serving 50 farmers. The movement has grown until there are now 29 clubs employing 33 advisers for approximately 1,460 farmers.

The main aim of the club is to improve economically the production of its members' farms. Each club has a membership of 50 and is managed by an executive committee comprising a chairman and five or six members. Each member pays an annual fee of from £35 to £50, depending on the size of his farm. The clubs usually employ a public accountant (as well as an adviser) to act as secretary to the club. An adviser is expected to visit each member at least once in two months. Some clubs buy farm needs in bulk to get discounts, the benefit of which is passed on to members.

Monitor farms in New Zealand were established in the 1980s and are described by Ian Riddle – who helped to transfer the idea to Scotland in 2004 – as 'groups [that] are farmer owned community groups consisting of from 25–70 members who chose one of their number as a central monitor farm for a three year period. The group is assisted by a facilitator and other specialists to monitor and measure, and to identify steps which can be taken to improve physical and financial performance. The monitor farm is used as an example to motivate others and speed up the improvement of farm performance within a local area'. Furthermore, 'the choice of monitor farmer is more important than type of farm. He should be of above average ability, willing to open his farm to others and accept policy decisions from the community group. Meat NZ now aim for farmers who are in the "top 20 per cent" in order to increase the rate of change. Privately organised monitor farm groups have been formed consisting of selected "top performers" who want to improve at a faster pace' (Riddell, 2001).

Monitor farms are very innovative examples of benchmarking for best practice, and through experimentation and group decision making are a rare example of benchmarking that leads to generative as well as adaptive innovation. In New Zealand, a report by Meat NZ found that 'farmers were asked what financial benefit they received per year as a direct result of their involvement with the MFP [monitor farm project], due to the adoption of the four technologies studied for each MFP. The results were grossed up on a stock unit basis to calculate an average net benefit per farm for each area. The overall average for the 90 responding farms was an annual net benefit of $6,700, or approximately $1.50/su per year for each farm' (Garland and Baker, 1998 Executive Summary).

In Scotland, where the groups are only in their first few years of operation, one group commented, 'The Community Group is working well and there is plenty of active discussion and debate at Monitor Farm meetings. Some of the work and recommendations by the group in the first year are now coming back through for observation and comments – for example, the reseeding trials, veterinary information and livestock diets' (Campbell, 2008).

INFORMATION RICH FARMERS

A study by Verissimo and Woodford (2005) from Lincoln University looked at the use of information by six of the top performing sheep and Beef Farmers in the South Island of New Zealand. Their conclusion was that top performing farmers are 'information rich'. 'The farmers used a broad range of strategies to acquire and filter information, and all were interested in benchmarking. This benchmarking comprised both formal and informal analyses to compare their performance with other top performing properties. They were not interested in how they compared relative to average performers, instead they wanted to compare themselves only to the best. They also had well developed processes for analysing and reflecting on what they read and saw, and then for testing strategies on their own properties.' Four of the six belonged to discussion groups, and the other two had been members in the past. Farm visits were an important source of information, either as organized field days or more informally, such as through sales and purchases of livestock. Some had travelled widely and visited farms overseas (such as South America).

Benchmarking was most readily done against local physical benchmarks and target setting was also on physical measures – lambing percentages, say, or target sale weights for animals. Comparative financial data was prepared by the accountant that all the farmers used, showing their performance relative to other clients. What was not benchmarked was managerial behaviour, such as information gathering techniques or planning, implementation and monitoring of operations, 'Yet,' the authors conclude, 'it is these managerial behaviours that characterise these farmers and which other farmers may wish to emulate.'

In terms of change, Verissimo and Woodford (2005) identify what is best described as an adaptive approach to innovation which is explained thus:

None of these farmers could be described as 'first movers' but all were early adopters. Having seen what other people were doing, and actively looking for innovative practices being undertaken on these farms, they would then assess how the practice might work on their own farm. But rather than making an impulsive decision they would 'brew it over' in their minds before deciding to act. Once having decided that the innovation had overall merit, they would typically test it on a small part of their farm before making any major commitment that could not be easily reversed.

Diffusion of the New Zealand Model

The New Zealand experience has diffused to other countries. Australia had management accounting groups from the mid-1960s sharing benchmark data, but discussion groups took off in the mid-1980s, following the pressures on the industry with the removal of production subsidies. Campbell et al. (1996) report that from the mid-1980s in Australia 'the majority of farmers have become involved with one or more groups to address landcare, productivity and farm business management issues'. The authors worked with two umbrella groups – Farm Management 500 and Farm Vision 2000 – to establish the benefits and drawbacks of farm discussion groups. Farm Management 500 is a private extension company that in 2008 stated it was working with 320 families across Southern Australia, with 30 groups in operation and maintaining a benchmarking database of 200 farms. Campbell et al. (1996) made an important observation about the value of farm discussion groups to themselves and others by saying:

At times farmers suffer from a feeling of professional isolation. They are most vulnerable to these feelings during periods of unprecedented change. This has been the case in the last ten to fifteen years. Our involvement with groups, has provided us individually with the support of others who share a belief that change, although uncomfortable and threatening at times, can also be a source of opportunity.

They also acknowledge that the longevity and success of such groups is due to the facilitators and advisors that they have:

Agronomists, both independent and agribusiness employed, are frequently asked for their opinions on a wide variety of production issues. In fact, the simultaneous emergence of groups and the ready availability of agronomic advice have complemented each other in bringing about beneficial change. However, the real value of groups is that because of our involvement in them we are better able to tap into the expertise and support of large companies, government departments and other progressive farmers. As a result, we now enjoy access to a wide network of colleagues, many of whom qualify as friends. The strength derived through our interactions within these networks has given us the confidence to move forward, to test out and try new techniques, to learn, to move on.

Farm management clubs have also been established in Canada since the 1968 and are particularly strong in Quebec and Ontario. In a presentation to the 2003 Managing Excellence in Agriculture Conference, the Secretary-General of the Fédération des Groupes Conseils Agricoles du Quebec, makes it clear that these groups are more than discussion groups and operate in a similar way to the Farm Bureaus of the US. They have administrative committees and salaried personnel, and offer a range of agronomy and business advice. More recently, the concept of discussion groups has been experimented with but the experience of these groups is yet to be published. The UK experience with expanding farm discussion groups is the subject of Chapter 6, and the mixed experiences in the US are discussed later in this chapter.

How Improvement Groups and Clubs Function

Discussion groups develop their own ways of operating but there is a discernible pattern to the way in which they are conducted. Groups usually consist of between 10 and 20 members, and a facilitator. Facilitators are drawn from the ranks of advisors and consultants, usually privately employed, but also from university extension and advisory services. Core activities are monthly meetings (or between 9 and 12 meetings a year) where typically the facilitator talks through the analysis of the data provided by the farmers. These may be supplemented by visits to each of the businesses in the group in turn. Variations include holding training days, talks by invited speakers and conferences in addition to regular activities. Conferences occur where a consultancy or levy board supports a number of groups and brings them together. Some groups meet for discussions and site visits, but do not collect financial data.

MINI CASE STUDY: WEST COUNTRY DAIRY FARMER

A farmer from a family farm in the West of England with 300 acres and 240 cows explains how they came to take on New Zealand block calving methods on their farm and how benchmarking through discussion groups works in practice on the farm.

'The whole industry in New Zealand is seasonal because 95 per cent of their milk is dried and exported as powder, so the whole industry really is seasonal, they just all farm for the lowest cost of production; whereas in this country we've got such a high population we need a certain liquid milk supply. It was the groups in South West Wales that were the first to really look at block calving. For financial reasons, they were the first that were really under pressure as farm gate milk prices dropped and they had to look at ways that long term they were going to have a profitable business. I guess some of the producers in South West Wales involved with a New Zealand consultant actually looked at matching milk production to grass growth. If you match milk production with grass growth, you end up with a spring calving system. So the cows start calving six weeks before grass really starts rocketing, and then their milk is peaking when the grass growth is peaking, which is May, early June.'

'If you can do it like that then your feed costs are going to be minimal: we don't really purchase any feed, we don't feed any concentrates to our cows – that's a massive cost saving. Our yield per cow is low, very low in terms of industry standards, but the margin's there. The Kiwi consultant took us over to South Wales to see what these guys were doing who were mainly one or two years ahead of what we were doing. When I saw that some of them were actually going to do block calving, some of them had actually sold their whole herds and then just bought in cows that calved in the spring. Some of them took two, three years to change the calving pattern of every cow, and that's actually what we did.'

'Of course, because I'd worked in New Zealand I'd seen what a simple efficient system it was and, once I saw the guys in Wales doing it, I thought there was no reason we can't do it here. And of course it's a bit of a snowball effect; once you see someone doing it and making a success – and of course we've always shared all of our figures as well.'

'We're benchmarking very much against each other in the regional groups. To some degree we're benchmarking nationally because the full-time consultant is actually going to all these groups. He's picking up the data, analysing it and then he does put our group averages against the other group averages. You've got people who are on a similar system to you, and we put our own cost of production on a pence per litre basis, and also £s per hectare, and it's quite interesting. You can see where you're doing better and where you're quite strong, and also where you're quite weak as well. And then you can look at the farms in your group – because there are no secrets, you look across and Such and Such has got very low plant and machinery costs. We'll organize a visit to his farm just to talk about plant and machinery and really get into it and look at how these costs were made up and how does he get them so low. We still go every year to South West Wales to see how they're getting on.'

(As told to the author)

A survey of UK farmers in 2006 commissioned by the Food Chain Centre found that 17.5 per cent of farmers used some form of benchmarking, and of this number 56.2 per cent were receiving support via a formal service or group. Another 15.4 per cent of the total number of farms surveyed claimed that they benchmarked 'informally' by speaking to friends and colleagues about best practice, attending meetings, using consultants and using figures in the trade press. Dairy and pig farmers are most likely to benchmark, which is unsurprising given the nature of those enterprises and the record keeping involved. These enterprises are characterized by regularity of operations and day-to-day movements in financial and physical data, that translate to monthly and annual figures for monitoring purposes, and more like-for-like operations and farms are available for benchmarking. The survey also found that larger farms (defined in the UK as being over 300 hectares, or 741 acres) are more likely to be involved in benchmarking but also that some very small holdings are used in benchmarking. Of those surveyed, 67.3 per cent had been involved for more than four years, which tallies with the promotion and funding of business groups by the UK government following the so-called Curry Report in 2000.

There had been groups in the UK run along the lines of those in New Zealand from the mid-1990s and there have been purchasing groups and other meetings from before then, but 2001–2 was a significant point in widening the availability of business groups. The majority of groups in the survey met 1–3 times per year, with 21 per cent meeting 4–6 times and 19.8 per cent meeting 10–12 times. The main sources of information were a college or university, consultants or the press.

The most common types of working groups attended were discussion groups organized by an advisory group such as the Red Meat Industry Forum (RMIF) and those organized by a private consultant for groups of their own clients. The groups organized by the RMIF, MDB and others have central comparative analysis databases, like those discussed in Chapter 2, into which members feed their information. These databases can be filtered to provide information just for a group, providing analysis of members against each other, and against national averages/quartiles. The consultants' groups run in a similar way, but using client databases and sometimes data from national surveys. Data is input through standard forms, either by the individual farmer or secretary, or via the consultant.

Benefits of benchmarking, as perceived by survey respondents, were firstly a better understanding of costs (reported by 59.1 per cent). Improved working practices were seen both as a benefit and a key result, as was a better understanding of business drivers. Increased returns, improved quality of produce and improved customer service were other benefits. Other respondents reported that new business opportunities had arisen through discussions and that the group provided advice and support in dealing with the issues and paperwork arising from the reforms of the Common Agricultural Policy.

Of those who did not benchmark either formally or informally, the most common reasons given were that they were unaware of the practices or unconvinced of the benefits. A few who gave more active reasons for non-participation said that their own business was unlike others, and so benchmarking was unsuitable, or that it was too difficult or expensive. For the former, it is likely that benchmarking is still perceived as a financially based, whole farm or at least whole enterprise activity and that the potential of competitive, external benchmarking against other businesses in other sectors has not been explored, an aspect that will be discussed in Chapter 9.

Why Farmers Join Groups and the Problems that Arise

The following stories from the US, where farmer-to-farmer peer groups or learning groups are less well established, indicate some of the behaviours that can be exhibited within such groups. Firstly, there is a negative story from the *National Hog Farmer* news magazine in the US:

> *Peer groups can be helpful, insightful and supportive. In fact, a good peer group can improve the business success of all those involved.*
>
> *One such group ... includes 15 producers from Georgia to Nebraska and Texas to Minnestota. The impetus for the group was that they all were clients of the same swine consulting group.*
>
> *[They] met dozens of times, discussing health concerns, nutrition, buildings, breeding, equipment ... They learned from each other's successes and failures, shared their hopes and goals, and in fact, knew just about everything there was to know about each other's hog operations.*
>
> *Then they decided to put their production and financial numbers into a database to develop some benchmarks. They planned to compare the production and economic performance of their hog operations and learn from each other in this way, too.*
>
> *Truth is, it just didn't work out as they'd hoped.*
>
> *It seemed like such a practical idea ... we wanted to establish some benchmarks so we could track progress and compare ourselves with other producers. But when it came down to it, I guess*

the numbers we'd like to share were a little too personal. It was really like exposing every private part of the business to the rest of the group.

Not only were the group's numbers incomplete, when compared to production and financial data from other sources, [they also] just didn't make sense.

A mass participation benchmarking scheme might be more successful than ... peer group experience because of the ability of the participant's to remain anonymous. 'No one wants their fellow producers to see them as inadequate, so if the numbers can be traced back to an individual operation, producers may tend to exaggerate them. Let it or not, you could become a member of a liar's club'

(Lane, 1999)

A spokesperson from the University of Minnesota made the following comments whilst promoting the concept of farmer learning groups in the US based on his experiences in Australia:

The participants in that group were able to ask specific questions of each other about how they achieved given production levels or how they contained costs on their farms. This also provided benchmarks against which each farm could measure its own progress. Those farmers didn't view each other as competitors, but rather as cooperators from whom each could learn something of value.

Not a lot of U.S. farmers are willing to be that open about their farm management and performance. They tend to be more secretive and almost view the neighbor as someone they need to out-produce rather than someone from whom they can learn. There are exceptions, but they aren't real common here in Minnesota.

(Schwartau, 2007)

The key to success in small group benchmarking is therefore trust and a sense of collaboration. The willingness to share data, to keep good records and the social factor – groups meet on each other's farms, in public houses and in other social spaces – make learning through these groups potentially worthwhile and enjoyable. They also appeal to the ingrained habit of farmers to 'peer over their neighbour's fences' as a means of judging the performance of their own farm against that of others.

Drawbacks include the cost of setting up and running the group. Evidence from consultants interviewed in New Zealand suggested that by having to contribute to the costs of running a group, farmers had more ownership – and were consequently more demanding in asking for information, as well as more innovative in the type of information that they were asking for. Competiveness between group members could become an issue (with a certain amount of scepticism being added to the interpretation of numbers) but there were also some who recognized that each group had a lifecycle and that moving to a new group with more like partners became essential over time. However, the majority of groups appear to be geographically based and few examples are found of 'virtual' or widespread groups.

Criticisms of Small Group Benchmarking

The most vocal critics of benchmarking are also based in Australia and this is unsurprising as their main criticism is that what is promoted as benchmarking in agriculture is just another version of comparative analysis. Ronan and Cleary (2000) examined this debate and concluded that there is a need to distinguish between the practices of whole farm analysis and the use of financial ratios as targets and best practice process benchmarking.

Much of what passes as benchmarking, they observe, is small-scale comparative analysis. This provides farmers with profit figures, ratios and quartiles or deciles for decision making. Such small group analysis lacks the statistical validity of the national surveys conducted in Australia by ABARE or in the UK through the Farm Business Survey (though there are concerns about self-selection of participants in these surveys), or the State-wide databases compiled by the land grant universities in the US. Ronan and Cleary (2000) also contend that 'it is a fallacy that whole farm analysis can yield enterprise-based best practice benchmarks and that enterprise analysis can yield whole business outcomes'. There are too many options open to farmers, too many different types of farm and farmer to identify a 'best practice business structure' or 'best practice asset deployment'.

Best practice benchmarking in agriculture should address different goals to these. Ronan and Cleary (2007) identify product cost competitiveness, process costing and the use of 'social' data as the basis for best practice benchmarking. The best discussion groups are certainly making use of all these practices, but the majority are still at a learning stage, knowing their costs and making incremental changes. The potential to extend the 'benchmarking for better practice' offered by discussion groups and monitor farms to become 'benchmarking for best practice' is the subject of Chapter 9 of this book.

Another criticism is that the level of record keeping, involving both financial and physical records across the industry, is not sufficiently good to provide comparative information. One farmer interviewed, who belonged to a number of different clubs related to different enterprises (for sheep, arable and cattle) commented that he had to submit accounting data in three different formats. The Canadian Farm Business Management Council (2000) commented that one of the pitfalls facing benchmarking groups was 'trying to set benchmarks without having standardized the accounting reporting practices within the group'. The DairyBase project featured in Chapter 4 tries to overcome this by getting the farmers' accountants to submit data in standardized form. This is an ongoing problem. However, belonging to a business club is an incentive to both improve and to gain greater understanding of the numbers being recorded.

Sustainable Change and Benchmarking for Better Practice

What is clear from the studies of small group benchmarking in New Zealand and other countries is that it does create conditions for learning about how one's own farm and others work, about the costs that are higher than they need to be and about tactical and operational changes on the farm. If the practice of comparative analysis appears statistically flawed on a small level, this is compensated for by the knowledge and understanding within the group of 'what doesn't look right' and the trust within some groups that allows members to challenge each other. Membership of a group can give the confidence to make changes, based on the experience of others, and to experiment with new innovations. Taking the framework from Bogan and English (1994) referred to in Chapter 1, small group benchmarking meets the basic criteria for the stimulation of sustainable change in that the activities of the group can:

- create motivation for change;
- provide a vision for what an organization can look like after change;
- provide data, evidence and success stories for inspiring change;
- identify best practices for how to manage change;
- create a baseline or yardstick by which to evaluate the impact of earlier changes.

The farm in the mini case study in this chapter made wholesale changes to the farm initially by moving over to block calving from year-round calving. The resultant increase in profitability made it possible to maintain two families (father and son) rather than one as before, and continued attendance at the group inspired them to make continuous improvements to their patterns of working. A series of case studies published by the Food Chain Centre in the UK gives a number of success stories in this area. Comments about small group benchmarking for another dairy farm included:

benchmarking demonstrated that the core business structure was sound, and gave them the confidence to expand the herd from about 140 cows about 4/5 years ago to 180 in 2005, and they intend to reach 200 cows in 2006

through the benchmarking [the farmer] was able to see the benefit of feeding higher quality feed, with the higher returns that others in the group were achieving ... by concentrating on this area he has seen a 2 pence per litre improvement in his milk quality

finance costs are lower at the bank because he is providing them with better figures

[the farmer] enjoys meeting every month or so with the other farm members to discuss ways in which they can improve their farms, and also likes seeing the figures and how he is improving. It has given him personal confidence in the way in which he is running his farm, and the direction he wants to take the whole business

(Food Chain Centre, 2005)

These comments indicate that small group benchmarking can achieve sustainable, incremental changes and promote continual improvement in farming. As Ronan and Cleary (2000) say:

The acid test of a benchmarking activity is if it facilitates change for the better in how farmers do things and the results obtained; that is, sustainable productivity growth and better profits ... Without appropriate benchmarking activities, rural industries will be less able to react proactively and appropriately to these challenges ... the issues of industry benchmarking standards, farmer participation and systems evaluation are central challenges.

Chapter 6 contains an account of how government funding in the UK was used to promote the use of business clubs in the period 2002–5, and the challenges of getting widespread take-up for the idea. It is believed that in New Zealand and Australia 51 per cent of farmers belong to such a group (Schwartau, 2007) making them the 'information rich' farmers that are more likely to make changes that are sustainable in business terms (Verissimo and Woodford, 2005).

References

Bogan, C.E. and English, M.J. (1994), *Benchmarking for Best Practices: Winning Through Innovative Adaption* (New York: McGraw-Hill).

Camp, R.C. (1989), *Benchmarking – The Search for Industry Best Practices that Lead to Superior Performance* (Portland, Oregon: ASQC Quality Press).

Campbell, N. (2008), *North Argyll Monitor Farm Ardachy Farm Year Two*. Annual Report. (Oban: SAC/ Scottish Government/QMS) <http://www.qmscotland.co.uk/analysis/downloads/year%202%20 annual%20report.pdf>.

Campbell, T., Bales, N. and Postlethwaite. N. (1996), Farmer discussion groups – past, present and future, *Proceedings of the Australian Agronomy Conference, Australian Society of Agronomy* <http:// www.regional.org.au/au/asa/1996/contributed/116campbell.htm#TopOfPage>.

CFBMC (2000), *Farm Management Clubs: A Road Map* (Ontario: Canadian Farm Business Management Council).

Edwards, J. (1996), *The Role of the Farm Office and the Provision of Administration in Various Countries: A Study Tour of Australia and Three Countries of the European Union*. National Farming Scholarships Report 1996 (Uckfield: National Farming Scholarships Trust Secretariat).

Fettes, M. (2006), *Banff & Buchan Monitor Farm Project Annual Report 2006* (Edinburgh: SAC Publications).

Food Chain Centre (2005), *Benchmarking through Business Clubs* (Watford: IGD).

Garland, C. and Baker, D. (1998), *Meat New Zealand WoolPro Monitor Farm Programme Cost Benefit Analysis* (Masterton: Meat NZ).

Grusenmeyer, D. and Sheils, C.M. (2004), The Power of Farmer-to-Farmer Discussion Groups, <http://media.cce.cornell.edu/hosts/agfoodcommunity/powerf2fdiscgrps.html>.

Kyrö, P. (2003), Revising the concept and forms of benchmarking, *Benchmarking: An International Journal* Vol. 10 No. 3, 210–225.

Lane, N. (1999), Benchmarking without baring your business soul, *National Hog Farmer* 15 August, 1999.

Riddell, I. (2001), *Monitor Farms and Farmer Discussion Groups In New Zealand (Summary)* (Scotland: A Farmers Club Charitable Trust Scholarship Report). <http://www.sac.ac.uk/mainrep/pdfs/ monitorfarmnz.pdf>.

Ronan, G. and Cleary, G. (2000), Best practice benchmarking in Australia: issues and challenges, *Agribusiness Perspectives* 2000 <http://www.agrifood.info/Review/Perspectives/2000_ Ronan/2000Ronan.htm>.

Schwartau, C. (2007), Build your farm a management advisory team with a dairy discussion group, University of Minnesota Extension <http://www.extension.umn.edu/dairy/dairystar/06-09-07- Schwartau.htm>.

Verissimo, A. and Woodford, K. (2005), Top performing farmers are information rich: case studies of sheep and cattle farmers in the South Island of New Zealand. Published in the Proceedings of the Fifteenth International Farm Management Association Congress, Campinas, Brazil, August 2005, Vol. 1, 365–368.

6 Farm Benchmarking in the UK: An Assessment by the Food Chain Centre, 2002–7

PETER WHITEHEAD
IGD

Introduction

In August 2001 Sir Don Curry was appointed by the government to lead a Policy Commission on the Future of Farming and Food. The Commission reported in 2002 with wide ranging recommendations, including the establishment of a new Food Chain Centre at IGD with a role that included to 'act as a champion for farm benchmarking'.

The Policy Commission believed that many more farmers could benefit from analysing their costs of production and comparing this with their competitors at home and abroad. The Commission highlighted considerable differences between the highest and lowest cost producers that could not be explained by geography or farming system. As a result they called for a stronger drive on benchmarking.

Farm benchmarking therefore became a core activity for the Food Chain Centre. Between 2004 and 2007 the Centre ran a programme aiming to increase the uptake and impact of farm benchmarking. This started by understanding the current position and promoting awareness of the benefits of benchmarking. From 2003 the Centre also managed the establishment of business clubs among fresh produce growers through a service called Hortbench.

In this chapter we will explain:

- how farm benchmarking developed over the period;
- our vision for farm benchmarking;
- why farmers benchmark;
- the critical success factors needed for further progress;
- our recommendations for future work.

Existing Situation

Benchmarking is a tried and tested business improvement method. Xerox is often credited with having initiated the modern era of benchmarking in the late 1970s when it decided systematically to compare its US processes and product costs with those of its Japanese affiliate. This work demonstrated the power of data comparison and today the approach is widely used by businesses in all sectors of the economy.

At the start of the Food Chain Centre's work the current position on farm benchmarking was unclear. The first steps involved:

- inviting leading experts to workshops at the Food Chain Centre to discuss benchmarking and to identify all of the services available for farming and food businesses;
- interviewing most of the leading farm benchmarking service providers;
- commissioning a telephone survey of 1,200 farmers and growers to establish their experiences of benchmarking.

At the time we found that a variety of benchmarking services were available covering all of the main sectors of farming, run by a variety of organizations including consultants, accountants, banks and others. These involved a wide range of approaches, from self-help to supported applications.

We were able to establish that a minority of farmers (around 8 per cent) were applying benchmarking, some for many years, and believed that it helped them improve farm profitability. We were encouraged to find that for these farmers, benchmarking really did work and they were committed to it.

The experts told us that benchmarking was not an end in itself but a means of promoting improvement for the farm. We saw that the best delivery mechanism was through local groups of farmers organized in 'business clubs' because this encouraged rigour and high-quality discussion.

However, this initial work clearly showed that there was nothing like a critical mass of farmers involved. The main stumbling block was not a lack of interest in benchmarking but a lack of awareness, particularly of how to get started. To address this, the Food Chain Centre produced a 'starter pack' for farmers that set out an explanation of farm benchmarking together with information on the services available. The starter pack was distributed to many hundreds of farmers and subsequently updated. The pack can be downloaded from www.foodchaincentre.com.

This work provided the foundation for more in-depth activity from 2004 onwards.

Vision

From the start we believed that farm benchmarking could make a real difference to profitability and we wanted to encourage the highest possible standards for service provision. The evidence available suggested that farm benchmarking was likely to be most effective when:

- it is organized around business clubs led by a trained facilitator;

- the clubs use financial data (as well as physical data) to identify differences in the true cost of production;
- comparisons take place between clubs locally, regionally, nationally and where possible internationally;
- improvement plans are generated and implemented based on benchmarking results;
- the process is repeated at regular cycles.

This vision would not be easy to achieve in practice. For everything to come together there has to be a willingness among farmers to share information. The necessary infrastructure also has to be in place, with sufficient quality facilitators. The schemes must guarantee the agreed levels of confidentiality to participating farmers.

Although our research demonstrated that there were strong farm benchmarking services in place when the Policy Commission reported, they fell short of best practice.

Improvements have taken place over the last five years; in particular the availability of benchmarking services has been transformed. These developments were led by the Levy Boards in particular and by the Food Chain Centre.

Sector specific schemes have been introduced covering all the major farming products, as shown in Table 6.1.

These schemes provide:

- bespoke benchmarking templates that are enterprise specific;
- the ability to capture production and financial data that enables the true costs of production to be identified;
- a reconciliation of benchmarking data to financial accounts so it is based on real-time information;
- a common platform run by Prospect Management Services known as Prime Numbers that provides comparison flexibility across groups and boundaries;
- web-based inputing and reporting.

Global research on farm benchmarking practices conducted in 2006 and summarized in Chapter 3 of this book indicates that this infrastructure is world-leading.

The new services have enabled some 230 new clubs to become established in England since 2003. Typically each club has around 6–10 participant farmers/growers. These clubs are in addition to those run by private organizations, which have also enjoyed growth over the same period.

Table 6.1   Sector involvement in benchmarking for business improvement groups

Sector	Organization
Arable farmers (CropBench)	Cereals Industry Forum
Dairy farmers (MilkBench)	Milk Development Council
Fresh produce growers (HortBench)	Food Chain Centre
Potato growers	British Potato Council
Red meat farmers	Red Meat Industry Forum

The infrastructure has thereby been put in place giving a major boost to farm benchmarking. Although this represents a major step change, further work is required to realize the full potential because subsequent surveys have shown that the majority of farm benchmarking is still informal.

Few, if any, farm business clubs have yet fully achieved the vision though many strive to. For example, many have to start with benchmarking production data to build confidence among their participants and so far there has been little data sharing across national boundaries, though the International Farm Comparison Network (IFCN) has well-established procedures for such work.

Benchmarking has not yet reached a critical mass of farmers in all sectors and there remain many barriers to participation that will only be overcome when there is a more widespread endorsement of collaboration and when a greater number of farmers have become fully versed in analysing financial data.

Tracking Participation

From 2004 the Food Chain Centre promoted farm benchmarking across all farming sectors through workshops and in various other ways. The Centre delivered some 44 events attended by 686 farmers. The Levy Boards, who have led the introduction of farm business clubs outside the fresh produce sector, have reinforced this through promotional efforts within their own particular sectors.

As a result there has been a great deal of coverage of farm benchmarking in the trade press, helping to raise its profile. The availability of farm benchmarking services coupled with promotional effort by the Levy Boards and others has delivered the significant increase in farm participation called for by the Policy Commission.

Since 2002 the Centre conducted two further surveys using the same methodology and the same database to track participation in farm benchmarking. The data in Figure 6.1 shows increasing participation from 8 per cent to 33 per cent in all types of farm benchmarking and from 2 per cent to 10 per cent in farm business clubs.

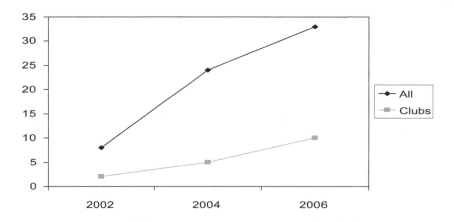

Figure 6.1   Percentage of farmers involved in benchmarking and improvement groups or clubs

Source: Food Chain Centre.

Although this shows a substantial improvement, it also demonstrates that the preferred form of benchmarking (through farm business clubs) is still in its infancy.

The research also shows there are substantial differences in participation between the main sectors of agriculture, as Figure 6.2 demonstrates.

Dairy farming has the best levels of uptake. The sector is well served, with benchmarking services offered by, for example, the Kingshay Trust, various consultants and the Milk Development Council. At the other extreme, levels of participation in the beef and sheep sector are below 10 per cent.

The evidence available suggests a higher uptake among farmers with above average acreage or herd sizes. Both the largest and the smallest size farms have the lowest participation. Although data is not available to confirm this, penetration by production, volume is likely to be higher than the figures on this chart.

There are good reasons for the differences in uptake by sector. For example, there are more small-scale operators in beef and sheep farming. Furthermore there was very little history of benchmarking in these sectors before the work of the Red Meat Industry Forum (RMIF) from 2003 onwards and so this level has been reached from a 'standing start'.

The RMIF and its partner EBLEX (English Beef and Lamb Executive) have recently embarked upon a scheme to boost participation from beef and sheep farmers using an approach known as 'Snapshot', which involves farmers completing a single-page form with basic production information. This is now being widely used.

Another reason for the difference in participation levels relates to production cycles. These vary from the daily production of milk to annual cropping and even longer rearing patterns. The greater regularity of data in the dairy sector has helped establish farm benchmarking as a more regular practice.

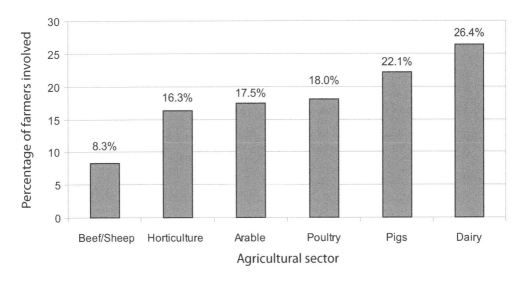

Figure 6.2   Percentage of farmers involved in benchmarking groups by sector

Source: Food Chain Centre Survey.

Why Farmers Benchmark

Farmers benchmark to compare their own farms with other farms of a similar size and structure. This process, the act of comparison and the issues which it highlights, helps them to plan and improve their businesses.

Over time the Centre's research shows that the benefits of farm benchmarking highlighted by farmers have changed (see Table 6.2). A rising proportion of those farmers benchmarking are reporting tangible commercial benefits.

These results suggest that the improvements made to benchmarking services over this period are paying off and that as more people gain experience, the benefits are escalating.

In the Centre's 2006 survey farmers identified their main reasons for benchmarking. The top five reasons in rank order are shown in Table 6.3.

Many farmers in business clubs have reported that the data derived from benchmarking has also helped them in negotiations with their supply chain partners. It enables them to build a better understanding of their own and others' breakeven points. Each reason for benchmarking is considered in turn below.

Table 6.2   Percentage of farmers indicating performance gains through benchmarking group involvement

	% Farmers	
	2002	**2006**
Benchmarking increased returns	4	34
Benchmarking helped understand costs	42	69

Source: Food Chain Centre Survey.

Table 6.3   Reasons for benchmarking by ranking

Reason for benchmarking	**Rank**
Understand costs	1
Improve working practices	2
Understand business drivers	3
Increase returns	4
Improve quality	5

Source: Food Chain Centre.

UNDERSTAND COSTS

All farm benchmarking schemes that have been introduced by the Levy Boards and the Food Chain Centre use a template that breaks down direct and indirect costs into several headings. In some cases there is a 'light' version where aggregate data only is compared but most also allow for a more detailed comparison of costs where these are available.

The templates are completed using data from farm accounts (as illustrated in Table 6.4). Because of different accounting conventions and different farm year-ends some care has to be exercised when transferring information from the accounts to the template.

We found time and again that farm accounts are not normally used as a tool for managing the business and that understanding of the underlying factors driving enterprise profitability is low. For some farmers there was a fear of finding out just how much they were losing on a particular enterprise.

Most farmers required support in order to understand all of the terms used in the accounts and interpret them correctly. In all the benchmarking schemes that have been introduced, a great deal of the effort has gone into developing the financial skills of the participants.

Through their business clubs, farmers are able to establish a starting benchmark for a run of between three and five years from historic farm accounts data. They are able to see how their costs have changed over the period and how well their costs compare with their peers.

Table 6.4   Sample spreadsheet: allocations of direct costs

Direct Costs	£	Cost allocation and basis		Calculated total	£/Ha
Seeds and plants	£25,000	20%	Overide %	£5,000	£50.00
Fertilizer	£12,500	30%	Automatic	£5,357	£53.57
Agrochemicals	£43,000	30%	Automatic	£18,429	£184.29
Agronomy	£15,000	15%	Automatic	£6,429	£64.29
Haulage	£2,500		Automatic	£1,071	£10.71
Crop Sundries	1,500		Automatic	£643	£6.43
Levies	£3,564		Automatic	£1,527	£15.27
Marketing	£2,851		Automatic	£1,222	£12.22
Packaging	–		Automatic	–	–
Other direct costs (non onion)	–				
Total Direct Costs	**£105,915**			**£39,678**	**£397**

IMPROVE WORKING PRACTICES

A benchmark of farm costs will highlight areas of strength and weakness. The reasons behind these differences can then be probed in discussion sessions to reveal those working practices that are giving lower cost producers the edge.

Farm business clubs with six or more members are likely to have cumulated experience of well over 100 years. Most members will be able to learn from each other. Even the best are unlikely to be good at everything.

Some benchmarking schemes are being linked to knowledge transfer programmes on technical research. As a result of benchmarking many farmers are making changes to the structure of their businesses and the way they work. This could include, for example, planting new varieties, reducing mortality levels, collaborating on the use of machinery and improving energy efficiency.

The benchmarking templates that have been developed provide for both direct and indirect costs to be attributed against an enterprise. One important consequence of this is that it has highlighted how the farm's indirect costs should be attributed to a particular operation.

For many farmers the control and management of indirect costs like electricity and management time have been highlighted as an opportunity for savings, a point missed by previous assessments based on 'gross margins'.

UNDERSTAND BUSINESS DRIVERS

Business clubs have increased farmers' awareness of factors beyond their costs that can improve their businesses. This potentially covers a wide range of issues, from market development and consumer trends through risk assessment to information and other technological changes.

INCREASE RETURNS

Some 34 per cent of farmers told us that they have increased their returns through benchmarking. Case studies that demonstrate improved returns from farm benchmarking can be found on the Food Chain Centre's website. Table 6.5 gives examples of farmers that are featured.

IMPROVE QUALITY

Farmers have been able to improve quality by focusing on better working practices. Improved quality generally also delivers increased returns.

Critical Success Factors

Great strides have been made over the last five years to put in place a world-class farm benchmarking infrastructure. Although this has bought about some highly desirable changes, alone it does not ensure success in the future. The main barriers we encountered are outlined below:

Table 6.5   Case study famers and gains through benchmarking

Case study	Description
Mary Quicke	Dairy farmer and cheesemaker used benchmarking to reduce her milk production costs by an exceptional 7.5 pence per litre.
Trevor Atkinson	Technical director helped Sentry Farms to reduce fertilizer costs by 10 per cent with no drop in yield, reduced repair costs worth £11/ha and cut labour and equipment costs by 12 per cent while increasing yield.
Alec Raffe	Dairy farmer has improved his milk quality and received an extra 2 pence per litre since he joined a business club in Cornwall. The group as a whole has seen feed costs fall by 10–25 per cent.
David Craig	Arable farmer from Shropshire reduced his chemical costs by 16 per cent.

- There remains a prevailing culture of independence among farmers, whereas business clubs require an ethos of sharing. This ethos is fully compatible with a very competitive approach to business. For many farmers the main competition comes from outside not within national boundaries. Also within a club framework, varying degrees of confidentiality are possible depending on the level of trust and openness of the participants.
- The business approach of many farmers naturally reflects the conditions they have grown used to, including the long-term subsidization of production. However, these conditions have been changing, and education and training provision need to adapt too. While this is evolving as a new generation of farmers come to the fore there needs to be greater emphasis on continued professional development for the established generation.
- Farmers faced falling incomes for a number of years and as a result have already had to reduce costs as far as possible. As a result, many were sceptical about the value of benchmarking. At the start it was unproven for the majority and farming was frequently regarded as uniquely different to other sectors where benchmarking had been applied. Furthermore, it was widely claimed by farmers that data from any benchmarking would be used against them and force down prices further.
- The recent rise of farm prices in some sectors could also discourage participation because it offers an easier route to profitability (although alone rising prices will not help farmers to maximize their profits).
- A great deal of time and effort in benchmarking continues to be taken up with what might be called 'advisory support'. This translates primarily into helping farmers analyse their accounts.
- There are numerous providers of farm accounts and they tend to have slightly different approaches including how they define items within the accounts. As a result there can be problems with comparing data; for example, there is no common approach to depreciation, and data has to be transferred manually to benchmarking templates. Further work is desirable to make it easier for farmers to populate benchmarking templates.

- There is also the issue of data capture for benchmarking and reporting results revolving around information technology. Not all farmers have an internet address and anyway some were reluctant to send financial data over the internet. Sometimes there were poor or non-existent broadband services, although this problem may soon be overcome.
- A lack of experience and confidence in using computers and in some cases a lack of a computer altogether was another underlying problem.
- Finally we have seen the difference that a good facilitator can make to the operation of a farm business club. Their work goes way beyond that of organizing and administering a meeting, collecting and inputting data. It requires, for example, an understanding of small group dynamics, and ways of learning and enthusing others to improve. Individual schemes developed their own solutions to facilitation but there was no recognized standard. The Food Chain Centre set about tackling this in conjunction with Writtle College and developed the first accredited training programme of its kind specifically for facilitating groups in agriculture.

Recommendations for Future Work

There has been a major drive on farm benchmarking since the Policy Commission report in 2002 and this has delivered very tangible commercial benefits. If ever there was any doubt that farm benchmarking could lead to improved profitability the evidence is now there conclusively to prove it. However, despite this success, farm benchmarking needs further attention to realize its full potential.

Over the period since 2002 work has been undertaken to address the natural scepticism among farmers. Benchmarking templates have been built, piloted and used extensively. But improvement plans and their implementation are still at an early stage. It takes rounds of benchmarking to iron out inconsistencies and establish trends. It takes even longer measured in crop or livestock cycles to see the effect of changes in working practices on farm profits.

The Food Chain Centre has made the following five recommendations:

1. Participation has not reached a critical mass of farmers. Business clubs should be continued and schemes like the RMIF 'Snapshot' developed in other sectors where there is a desire to reach smaller farms. For the more advanced, planning tools should be developed that can be used to evaluate production options.
2. Improving the ways that data is transferred between accounts and templates would make it much easier for farmers to participate in benchmarking schemes. This requires greater automation. It also implies that farm accountants could extend their role.
3. Business critical information other than finance, for example quality data, should be integrated in benchmarking schemes so that farmers can link these aspects better together.
4. Facilitator training programmes should be extended around the standard set by Writtle College.
5. Business club agendas should keep widening to equip their members with other critical skills, for example business planning and marketing.

PART 3

Environmental Benchmarking in Food and Farming

7 Standards, Indices and Targets for Environmental Performance

Introduction

The notions of sustainability and environmental protection have produced a significant amount of effort by government and researchers to identify and create performance measures and indicators, targets and indices. The natural foci for this activity are agriculture, being fundamentally land-based, and the food industries, which incorporate many features other than food production that impact on land through packaging and transport. Despite this effort and the plethora of models that have been produced, it is debatable whether benchmarks or benchmarking of environmental activity exist in the agri-food chain as defined in the general literature. The aim of this chapter is to identify a number of areas of performance measurement that have been labelled as environmental or sustainable in farming and food, and to assess whether this activity could be construed as benchmarking, or developed into benchmarking. Here, as throughout this book, the notion of benchmarking carries a meaning of using performance measures to drive operational and strategic changes in a business.

A significant amount of activity in this area has been led at policy level, but as seen in earlier chapters, the provision of benchmarks is not the same as benchmarking for best or better practices. Therefore, the policy side of identifying and establishing measures is not covered in this chapter. There are a number of publications that do cover the policy aspects of environmental and sustainability measurement for example Brouwer and Crabtree (1999) and Schulze (1999). This chapter does cover some basic principles of how environmental and sustainable policy has been worked into legislation, through nutrient budgeting, externalities and agri-environment schemes, and as these schemes have achieved success in changing farm practices, the potential of such schemes to be incorporated into benchmarking activities at farm level is evaluated.

Sustainability and environmental protection are complex, and by definition not completely predictable, issues. It is not possible to fully predict the consequences of certain types of intervention or action, and because it seems almost impossible to create methods of farming and food distribution that are either neutral in their effects on the environment or profitable (or preferably both), the issues are by necessity broken down into discrete problems. Thus, the issues of performance measurement and monitoring are broken down into:

- land/soil;
- water;
- air;
- waste;
- bio-diversity.

From there, problems of degradation of land, pollution of water, rising salinity levels, mineral levels and loss of wildlife species arise as being potentially measurable. These factors can not be measured without scientific expertise, and the need for expertise raises an immediate problem with setting benchmarks or absolute standards. Hence, much of the activity in this area to date has been concerned with setting standards and monitoring compliance with those standards. But standards tend to be set nationally or internationally, while farms work in local environments. However, as discussed below, compliance with schemes imposed from government has led to innovative changes in farming practice and in food distribution.

Four examples are used in this chapter to illustrate how issues of environment and sustainability have been tackled in the agri-food industry. These can be split into quantitative and qualitative approaches:

- Nutrient budgeting (or input–output accounting) is a quantitative method of approaching the problems of excessive nitrates and other minerals in soils.
- Agri-environment schemes, by contrast, are qualitative and require plans to be developed against desirable target activities geared towards preservation of landscape and wildlife. The use of demonstration farms using integrated farm systems is another activity that incorporates aspects of benchmarking.
- Certification schemes have proved to be a popular means of farmers, in particular, taking on additional environmental measures in their activities, and the standards for certification act as benchmark statements for agricultural practice.
- Codes of practice adopted by the food industry.

Whilst none of these examples is, according to the definitions used in this book, 'benchmarking for best practice', they all contain features of benchmarking that could be developed at farm level and within the food supply chain to contribute to sustainable change.

Benchmarking offers the opportunity for networked groups to develop their own targets and monitor local practices, to drive innovative changes in practice. The central question is whether voluntary action in this area is preferable to legislated action. In Chapter 8, Sue Kilpatrick of the University of Tasmania offers observations of four such groups in Australia, who voluntarily set and monitored different benchmarks relating to land and water to prevent degradation of those elements.

The final aspect of environmental protection in the industry is the link with payments and penalties from government to ensure compliance. Is this required to achieve sustainable change, and is it a factor in the development of benchmarking for best practice in this area?

Nutrient Budgeting: Using Policy to Set Standards and Monitor Compliance

There are a large number of schemes that have been developed at policy level that are aimed at reducing the levels of nitrates and other minerals in soil, water and air. Technically, these schemes require target and standard setting, and enforced compliance, rather than benchmarking. There are few, if any, examples of mass participation or small group benchmarking that use these measures as benchmarks for best practice (as distinct from performance indicators). However, there is a method of accounting associated with the control of agri-chemical inputs that is used in land-based businesses, namely 'input–output accounting' (IOA). This term is used in the EU, whilst 'nutrient budgeting' is the preferred term in Australia, New Zealand and the US.

DEFINING NUTRIENT BUDGETS/INPUT–OUTPUT ACCOUNTING

A nutrient budget can be defined as an accounting approach for nutrient inputs, stores and outputs. Although there are a range of approaches to nutrient budgeting for farms, they can be classified into three major types: farm-gate, field and farm-system budgets. The choice of a budgeting approach depends on the intended purpose of the study, which in itself should also define the scale, the required accuracy, the data required, and the data collection strategy. A farm-gate budget operates as a simple accounting of nutrient inputs and outputs, and integrates farm scale information into an environmental performance indicator.

A surplus or deficit can be adjusted for changes in stored soil nutrients and is often used to estimate loss, especially for nitrogen. Field budgets record nutrients that enter the soil surface and leave the soil via crop or pasture uptake. Field budgets are used for estimating the net loading of the soil with nutrients and can assist in refining soil nutrient distribution patterns within the farm. Farm-system budgets attempt to determine all nutrient inputs and outputs, transformations of nutrients within farm systems, and changes in soil nutrient pools. Farm-system budgets can be used to quantify nutrient loss pathways to water or air, either through direct measures or model predictions. These three budgeting approaches complement each other. Farm-gate budgets are most common as they are generally easy to calculate from readily available data at the farm scale and from sources that are likely to be fairly accurate.

(Extracted from Gourley et al., 2007)

The levels of nitrogen, phosphate and potassium (N, P and K respectively) have been of concern, particularly in livestock production, and form the basis of nutrient budgeting. Gourley et al. (2007: 1064) describe the problem succinctly: 'The risk of nutrient pollution from a dairy farm increases when nutrient inputs exceed the amount of nutrients leaving the farm in products. Total P and N inputs onto dairy farms, mainly in the forms of feed, fertilizer and N fixation by legumes, are usually much greater than the outputs of P and N in milk, animals and crops. These surpluses tend to increase as farms intensify and stocking rates increase.' The excess of nitrates also comes from manure re-applied to the land. As one review pointed out, manure is a cost-less fertilizer for farmers and so tends to be used liberally and to excess.

Legislation controlling the levels of nitrates and other minerals in the soil has been in place since the early 1990s. For example, the EEC Nitrates Directive in 1991 sought to limit the pollution of waters by nitrates leaching from agricultural land. Member States of the then European Community were required to put monitoring systems into place, and to identify national vulnerable zones (NVZ) where the application of nitrogen from animal manure was not to exceed 170 kg/ha. Member states were also required to set up codes of best practice. The problem in the early 1990s was not overstated. Brouwer and Hellegers in 1996 reported that:

> More than half (330,000) of the farms with a supply of animal manure exceeding 170 kg/ha are grazing livestock; the others being granivores (56,000 holdings) and mixed farms (144,000 holdings). The production of animal manure on about 30% of grazing livestock farms in the EU exceeds 170 kg/ha. This share is much higher at granivores (92%) and also high at mixed farms (21%). Differences among countries are largest at grazing livestock farms. Manure production exceeds 170 kg N/ha at more than 80% of the grazing livestock farms in Greece (83%) and the Netherlands (100%). The shares are lowest (below 20%) in Germany, France, Ireland and Luxemburg.

(Brouwer and Hellegers, 1996: 207)

They also expressed the view that major adjustments were required at such farms to meet the standards of the Directive. In Australia, Gourley et al. (2007: 1065) identify the intensification of dairy farming over the previous 25 years as the main reason for the increases in NPK in the environment, with most dairy farms being in net positive balances for NPK and in need of reduction. As elsewhere, nutrient budgeting or IOA is recommended as the mechanism to identify, monitor and promote innovation in the reduction of pollution and land degradation.

A survey of IOA systems in the EU and the US (Goodlass et al., 2003) identifies certain characteristics of these systems as operated. IOA systems are generally managed by agronomy firms and consultants, using software packages developed for the purpose. The development of packages and the costs of running pilot schemes are largely funded by government. Goodlass et al. report, 'All the nutrient systems gave farmers a specific explanation and this was usually in a written report. Comparisons with the farmer's own historic data or average farm values were the most common. Only ethical accounts for livestock farms (EALF) and agro-ecological indicators (AI) had any information on reproducibility or variation of results between seasons' (ibid. 20). They go on to say, 'Benefits in terms of increased awareness of problem areas were identified by several respondents. Anecdotal evidence suggests that farmers are encouraged to make actual changes to their management on the basis of the systems, if they receive detailed help from an adviser associated with the system, or if the system results in a marketing advantage.'

'It seems likely that IOAs could be used to increase awareness and provide evidence of the impact of management changes, they may need to be linked to supporting systems of technical advice' (ibid. 20). This implies that although there is a level of comparative analysis built into IOA systems it has not yet developed into a form of best practice benchmarking. One reason for this could be that farm business management advice and consultancy is offered separately from agronomy advice, and only where a holistic or whole farm approach is taken, where agronomists and business consultants work together, is it likely that best practice benchmarking might be integrated.

In Australia in 2007, Gourley et al. reported, 'Although there currently are fewer pressures or incentives within Australia to use nutrient budgets, and their use is low when compared with the EU, the USA or New Zealand, there is growing interest from advisors and farmers alike in using nutrient budgeting, particularly to assist with fertilizer management decisions. There is also growing interest from catchment management authorities and dairy companies, as nutrient budgeting is viewed as a useful tool in helping to achieve voluntary environmental nutrient management standards.' Commenting on the Goodlass et al. (2003) and Halberg et al. (2005) they say, 'Many of the tools identified were still in the research or pilot stage, with only a few having a high degree of acceptance and farmer use, while none are formally audited or linked to any marketing schemes' (Gourley et al., 2007: 1066). Goodlass et al. (2003) concluded that most of the tools and documentation did not provide assistance with interpreting the outputs, nor did they suggest farm management options for improvements.

The problem seems to be, as Melland et al. (2005) explain, that there are a large number of assessment tools that have been developed individually, and therefore there is no standardization or even consensus on what is required in this field. They go on to comment that 'selecting the most appropriate nutrient decision support tools presents a significant challenge for most farmers and advisors, and, therefore, farmers generally utilise tools that are locally developed and recommended by people they trust'.

Therefore, whilst the need to reduce nitrates and other elements in the soil is recognized and there exist numerous tools to enable farmers to carry out nutrient budgeting, voluntary take-up of such schemes is low. A higher degree of commitment exists where the schemes are being run by known advisors. Where there are advisors involved, then plans for improvement based on measured levels of nutrients against target or average figures are drawn up; what appears to be less common is the identification of processes and changes arising from comparison of one farm's activities against other farms in this area. Given that the data (even if flawed) is available, the next potential step would be to set up farmer-to-farmer benchmarking for best practice, as will be outlined in Chapter 9, or discussion groups, such as those trialled in the case study in Chapter 8.

Agri-evironmental schemes

Agri-environmental schemes were introduced by governments in European countries in the 1980s. The aim of such schemes is to provide incentives for farmers to engage in conservation and environmental protection practice, or as one website puts it, they are 'schemes that pay farmers to farm in an environmentally sensitive way'. Participation in agri-environmental schemes is voluntary. Core to agri-environmental activities is the preparation of a farm management plan, developed from a number of target activities encompassed in legislation, and the monitoring of compliance with plans.

The question of how benchmarking can be used in conjunction with agri-environmental schemes is interesting and a number of issues can be identified. There are standards, targets and performance indicators, but not systematic recording and monitoring accessible by farmers in the same way as the mass participation benchmarks discussed in Chapter 2. Participation in schemes is not universal and depends heavily on the attitude of the farmer to conservation and environmental issues. The schemes provide an income for farmers – one farmer claimed his agri-environmental scheme 'was another enterprise on the farm' – but whether farmers should be compensated for environmental protection actions is a matter of debate. The relevant questions here are whether agri-environmental schemes are successful in initiating changes in practice, and whether benchmarking for best practice could incorporate the requirements of agri-environmental schemes, or indeed, whether agri-environmental schemes could incorporate benchmarking.

Studies on the effectiveness of agri-environmental schemes broadly suggest that they are successful. Kleijn and Sutherland (2003) make the point, however, that there are very few robust studies of effectiveness: they look at evaluation studies relating to biodiversity in Europe and find a bias towards the UK and Netherlands, where only 6 per cent of the EU budget for agri-environmental schemes is available. Other studies have indicated increases in biodiversity and in bird populations. Primdahl et al. (2003: 137) report, 'Reductions in the use of nitrogen-fertiliser and pesticides were the most widespread type of obligations and significant effects of reductions of these factors were found in most of the case study area types and for most farm types. Significant effects, but less widespread, were found for indicators concerning livestock density, livestock maintenance, crop diversity and fallow land management.' Studies have also shown positive impacts on farm income (for example, Banks and Marsden, 2000), although Lowe et al. (1999) point out that income studies do not give indicators of success in additionality, or improvement to the delivery of public goods, that such schemes are also meant to deliver.

Despite the lack of comprehensive, reliable indicators, agri-environmental schemes are felt generally to be a success, although take-up is by no means universal. An interesting study was made into the participation in the Environmental Farm Planning scheme in Ontario, Canada in 2003 and into the endurance of changes made as a result of participation in the scheme. Smithers and Furman (2003) found that the scheme was regarded as innovative. Rather than requiring set targets to be met, farmers drew up a farm plan for engaging in conservation practices and were measured against their plan. Templates for the planning process are available for guidance, and guidance is available through a five-step programme (attendance at a training workshop; completion of self-assessment; development of a remedial action plan; submission of a completed farm plan for peer review and comment; application for financial aid in plan implementation). Participation was high – 17,000 participants by mid-2000 from a 1993 starting point – although not all participants completed the full five-step programme, with some producers completing plans and executing them, but forgoing peer review and funding. The most frequently addressed environmental issues were water contamination and soil degradation, issues that were already given attention through previous initiatives.

From the point of view of assessing the relevance of this type of study to the development of benchmarking in this area, the Smithers and Furman (2003) study identifies the following key factors:

- the motivation of farmers to take part in agri-environmental schemes;
- the farmer and farm attributes or type;
- farmers' perceptions of the programme itself;
- lack of definitive evidence for enduring change in farm practices.

The motivation of farmers to take part in any environmental change, particularly if there are no financial packages from government to support the changes, needs to be considered when advocating benchmarking in this area. Although participation in agri-environmental schemes is some indicator of success, it is inadequate to explain either the financial or business benefits. Pro forma planning templates, such as those used in the Ontario project or in the UK schemes, provide indicators of good practice but not necessarily best practice.

THE ROLE OF GROUPS AND ASSOCIATIONS IN AGRI-ENVIRONMENTAL SCHEMES

There are a number of groups in different countries that have been formed to promote environmental awareness and action involving farmers as well as other members of the community. These include Landcare Farming in Australia, which provides resources, grants and case studies of ways in which farmers can improve the quality of environmental care whilst retaining profitability. Members of Landcare are encouraged to form groups, and the organization provides resources for those setting up groups. The first Landcare Farming Group was established in 1986. Landcare claims to reach about 75 per cent of farmers in Australia (through 4,000 groups), and there is therefore a high degree of learning about environmental protection built into the infrastructure of farming. Landcare Farming promotes itself as 'a system of farming practices that are more compatible with the climatic constraints and land use capabilities of Australian landscapes', particularly the severe current problems arising from drought and high salinity levels (Cullen et al., 2003). Cullen et al.'s 2003 report on Landcare Farming also claims the following benefits for the groups:

Landcare group members also had significantly greater length of experience as farmers, due to the trend in the study area of retirees moving into farming for the first time. This study established that landcare participation made a significant difference to landholder awareness of issues, knowledge of land and water degradation processes, and to the adoption of recommended practices.

Later surveys by ABARE in 1995–96 ... found that compared to non-landcare group farmers, landcare group members:

- *operated larger farms with less intensive cropping and more livestock;*
- *recorded higher levels of farm income, farm debt and farm business capital;*
- *had a higher rate of return to farm business capital.*

(Cullen et al., 2003: 7)

Canada runs groups on similar lines, known as agri-environmental or 'green' clubs, where farmers can gain resources and information about environmental systems, such as pest control with reduced chemical inputs. An article from 2001 claims that there were then 75 green clubs in Quebec with around 4,000 members, and that data on integrated farming practices is pooled and used to monitor progress in the spread of integrated farm management in the state (Duchesne, 2001).

In the UK, as with Landcare and the Canadian groups, there are government-funded initiatives to bring best practice groups together in the area of agri-environmental care. The Food Industry Sustainability Strategy published in April 2006 by the Department for the Environment, Food and Rural Affairs (Defra) 'aims to improve the food industry's environmental, social and economic performance by encouraging the widespread adoption of best practice by the industry'. The report of the Food Industry Sustainability Strategy Champions' Group on water published in 2007 sets out an agenda and brings together current practice in the area. As well as promoting a number of initiatives and identifying sources of data, there is a case study based on the role of trade associations in promoting efficiencies and best practice. The British Beer and Pub Association (BBPA) operates what has been referred to in this book as 'mass participation benchmarking' by collecting data from its members (on a confidential basis) on the use of utilities and disseminating the results and best practice ideas through its website, conferences and workshops. The benefits are stated as:

The BBPA has found that this simple industry-wide initiative has been successful in:

- *establishing a data base for the industry;*
- *promoting energy and water saving;*
- *enabling member companies to benchmark energy and water usage;*
- *facilitating and enabling outside agencies to work with the industry on energy and water reduction programmes; and,*
- *promoting the industry as responsible in its use of energy and water.*

(FISS, 2007: 29)

To an extent, it can be said that best practice benchmarking is happening through such groups and that farmers do make sustainable changes to practices, as shown through the many case studies offered by Landcare or the UK group Envirowise, for example. Agri-environmental schemes and initiatives contain a number of elements that are identified as good practice in benchmarking, namely:

- self-assessment and long-term planning;
- sharing of case studies and resources through government sponsored agencies and privately run consultancies (such as Farm Management 500);
- data collection and target setting, albeit not on an agreed or systematic basis;
- opportunities for learning.

Weaknesses concern the attitudes of farmers and their expertise or willingness to engage in schemes. There seems to be a correlation, however, between the immediacy of environmental issues as a threat to the survival of the business – as in Australia – and the take-up of environmental schemes. In Europe, take-up is strongly related to the availability of funding rather than strong ethical or environmental concerns. It can, therefore, be broadly concluded that given that all the elements of benchmarking are available, practices are likely to become more developed as competitive advantage and business survival become more related to environmental action.

Certification Schemes

The standards used for certification schemes are classed by some people as benchmarks. They certainly fulfil the criteria of representing good practice or desirable characteristics which will enable or force the business or product seeking certification into making changes in practice to meet the standards. They do represent benchmark behaviours but are not benchmarking for best practice, in the manner that this term is understood in industry.

Standards which involve environmental practices have been judged to be successful in changing or enhancing farmer behaviours. Furthermore, the pressure to adopt standards and to apply for certification is promoted by the buying and promotion practices of wholesalers and retailers. The 'Soil Association' mark or the 'Demeter' mark show that certain practices are undertaken by the farms producing the goods being sold. Organic schemes are well established, but there are more recent schemes which seek to establish measures based on food miles or carbon footprint.

Another dimension is the use of demonstration farms, which essentially set benchmarks and experiment for best practice. The LEAF marque scheme in the UK exemplifies practice in this area (www.leafmarque.com). LEAF (Linking Environment and Farming) aims to work with all stakeholders in the food chain, including retailers, consumers, environmentalists, researchers and farmers (Drummond, 2006: 51). LEAF contributes to the notion of benchmarking in farming and food are through demonstration farms and innovation centres, and through its LEAF marque scheme. The ethos of the demonstration farms is to promote integrated farm management practices – 'a whole farm policy aimed to provide efficient and profitable production which is economically viable and environmentally responsible. It integrates beneficial natural processes into modern farming practices using the most appropriate technology' (ibid.). The farms are not only at the centre of research in the organization but are open to other LEAF farmers and to the public.

An article from *Farmers Weekly* illustrates the aims of LEAF:

Demonstration Farms play a critical role in demonstrating a living and working example of what can be achieved by adopting positive environmental practices,' said [the] Rural Development Minister.

They also provide a seeing is believing approach to the public by reconnecting consumers with what they eat and how it is produced – issues at the heart of LEAF.

[The latest Demonstration Farm] is an example of an integrated farming system showing how productive farming can coexist with environmental care.

Conservation activities include a programme of tree and shrub planting, a conservation reservoir and fencing off species-rich banks and verges.

We look forward to welcoming a range of visitors to the farm to discuss how food is produced and how the land is managed.

(*Farmers Weekly*, 19 October 2004).

The LEAF marque scheme offers certification (currently used in two UK supermarkets) against environmental criteria and these criteria also offer the chance for audit to take place both on the farm and through a computer-based planning system. Farm planning and assessment form a key component of the service, along with the opportunity to gain competitive advantage.

According to a Dutch study by Manhoudt in 2006 on the enhancement of biodiversity schemes on farms through environmental certification schemes 'The two main reasons cited by farmers for participating in a certification scheme were "to improve the image of farming" and "requests by retailers/supermarkets"... Most farmers were open to the idea of a certification scheme that includes criteria for pesticide use and nutrient use. These were regarded [as] more important than criteria related to biodiversity.' Manhoudt suggests that this is partly because (as discussed above) the Netherlands already has statutory schemes in place for these two activities.

Reasons for not joining environmental certification schemes (run by private or non-governmental organizations) were similar to those for not joining agri-environmental schemes, in that farmers did not want the restriction on practice, nor the increase in work (particularly paperwork) for the farm manager. Kragten and de Snoo (2003) suggest that a single system should be adopted that can provide all relevant information to retailers, governments and other parties, and observe that a number of writers have recognized that 'given farmers' attitudes towards participation in an environmental certification scheme, as described above, several actors in the agro-production chain, e.g. supermarkets or retailers, would seem to be very important for motivating farmers to participate'.

Codes of Practice and Standards

Retailers and supermarkets themselves have introduced codes of practice and targets which they aim to meet, and to which they aim to get suppliers to conform. Two examples here will give a flavour of how the wider food industry is addressing environmental issues. The Food and Drink Federation in the UK has issued what it terms its 'Five-fold Ambition' and this is given as an illustration of what codes of practice look like:

FDF members are committed to making a significant contribution to improving the environment by targeting priorities where they can make the biggest difference. Working collectively, our five-fold ambition is to:

- Achieve a 20 per cent absolute reduction in CO_2 emissions by 2010 compared to 1990 and to show leadership nationally and internationally by aspiring to a 30 per cent reduction in CO_2 emissions by 2020.
- Send zero food and packaging waste to landfill from 2015.
- Make a significant contribution to WRAP's work to achieve an absolute reduction (340,000 tonnes) in the level of packaging reaching households by 2010 compared

to 2005 and provide more advice to consumers on how best to recycle or otherwise recover used packaging.

• Achieve significant reductions in water use to help reduce stress on the nation's water supplies and contribute to an industry-wide absolute target to reduce water use by 20 per cent by 2020 compared to 2007.

• Embed environmental standards in our transport practices, including contracts with hauliers as they fall for renewal, to achieve fewer and friendlier food transport miles and contribute to an absolute target for the food chain to reduce its environmental and social impacts by 20 per cent by 2012 compared to 2002.

(www.fdf.org.uk)

In 2007, Wal-Mart introduced its 'Sustainability 360' plan which includes targets on waste, energy usage, diversity and its carbon footprint. This followed a year long 'self-assessment' process and although sceptical observers of Wal-Mart have pointed to its lack of time frames on targets, there is general interest in how well it manages to achieve its goals. If Wal-Mart can reduce the environmental impact of its many stores and lines of stock then others will follow – although, again, critics of Wal-Mart say they would like to see the same level of action applied to workers' rights. *The Guardian* newspaper in the UK (where Wal-Mart operates through the Asda chain of supermarkets) commented:

There are a handful of early signs that critics are being won over. Among Wal-Mart's projects is to cut back on packaging, which saves both money and waste. At Asda, salad bags are now 15% thinner and cardboard sleeves have been removed from a range of ready meals.

In September, Wal-Mart outlined a detailed 'scorecard' which it will use to mark the packaging processes of all 60,000 suppliers in an effort to encourage its entire purchase chain to go green.

In a rare outbreak of fraternity, the union-backed campaign group Wal-Mart Watch applauded the programme: 'We're encouraged that Wal-Mart has formalised its plans to reduce packaging and cut its carbon emissions. This goes to show Wal-Mart's power when it chooses to be a leader.'

(*The Guardian*, 6 November 2006)

The packaging scorecard is the most developed tool that Wal-Mart has produced and it is based on the premise that suppliers will be able to evaluate themselves against other suppliers. The metrics being used are, according to the Wal-Mart Press Release (2006), as follows:

• 15 per cent will be based on GHG/CO_2 per ton of production;
• 15 per cent will be based on material value;
• 15 per cent will be based on product/package ratio;
• 15 per cent will be based on cube utilization;
• 10 per cent will be based on transportation;
• 10 per cent will be based on recycled content;

- 10 per cent will be based on recovery value;
- 5 per cent will be based on renewable energy;
- 5 per cent will be based on innovation.

Suppliers will be given their own overall score and those supplier scores relative to it in each category.

There are a number of other codes and schemes that are employing benchmarks and targets to promote better practice and to enhance credibility of businesses in the industry. It is too early to assess the effectiveness of such codes of practice but, as seen earlier in the chapter, pressure from retailers and supermarkets to meet environmental standards is seen as a positive incentive to farmers and other smaller suppliers to put changes into effect to achieve environmentally beneficial targets.

Environmental Benchmarking for Sustainable Change

Rather than give a comprehensive account of all the approaches to environmental protection that have elements of benchmarking for best practice built into them, a number of illustrative examples have been given in this chapter. It is clear that sustainable change in the long run includes changes made to ensure sustainability of the environment, in particular soil, water and biodiversity. However, there is no straightforward approach to incorporating environmental factors into assessment and benchmarking. There is potential for treating environmental protection as a process and for applying process benchmarking techniques to establish best practice. One of the strengths of benchmarking should be taking an incremental approach to problems and aiming for continual improvement, rather than wholesale changes. This approach is explored in the Chapter 8, where four farm improvement groups adopt one different environmental measure and work together to improve practices. Further work is needed by farmers and their advisors to understand whether this is the best approach to take with changes for environmental reasons.

References

Banks, J. and Marsden, T. (2000), Integrating agri-environment policy, farming systems and rural development: Tir Cymen in Wales, *Sociologica Ruralis,* Vol. 40, No. 4, 466–480.

Brouwer, F. and Hellegers, P. (1996), The Nitrate Directive and farming practice in the European Union, *European Environment,* Vol. 6, 204–209.

Brouwer, F. and Crabtree, B. (eds) (1999), *Environmental Indicators and Agricultural Policy* (Wallingford: CABI).

Cullen, P., Williams, J. and Curtis, A. (2003), *Landcare Farming: Securing the Future for Australian Farming* (NSW: Landcare Australia). <http://www.landcareaustralia.com.au>.

De Snoo, G.R. (2006), Benchmarking the environmental performance of farms. *International Journal of Life Cycle Assessment,* Vol. 11, No. 1, 22–25.

Drummond, C. (2006), Integrated Farm Management Systems – making sustainability accessible to all: the LEAF viewpoint, *Aspects of Applied Biology,* 80, 2006, 51–63.

Duchesne, R-M., (2001), Strat{ea}gie Phytosanitaire, *La Fleuve* March 2001 Vol. 11 No. 10. <http://www.slv2000.qc.ca/bibliotheque/lefleuve/vol11no10/vol11_10_accueil_a.htm>.

FISS (2007), *Report of the Food Industry Sustainability Strategy Champions' Group on Water* (London: DEFRA).

Goodlass, G., Halberg, N. and Verschuur, G. (2003), Input output accounting systems in the European community: an appraisal of their usefulness in raising awareness of environmental problems, *European Journal Agronomy* Vol. 20, No. 17, 24.

Gourley et al. (2007), Nutrient budgeting as an approach to improving nutrient management on Australian dairy farms, *Australian Journal of Experimental Agriculture* Vol. 47, 1064–1074.

Halberg, N., Verschuur, G. and Goodlass, G. (2005), Farm level environmental indicators: are they useful? An overview of green accounting systems for European Farms, *Agriculture, Ecosystems and Environment* Vol. 105, 195–212.

Kleijn, D. and Sutherland, W.J. (2003), How effective are European agri-environment schemes in conserving and promoting biodiversity? *Journal of Applied Ecology* Vol. 40, 947–969.

Kragten S. and de Snoo G.R. (2003), Benchmarking farmer performance as an incentive for sustainable farming: environmental impacts of pesticides, *Comm. Applied Biological Science Ghent University*, Vol. 68 (4b), 855–864.

Lowe, P., Ward, N. and Potter, C. (1999), Attitudinal and institutional indicators for sustainable agriculture, in Brouwer, F. and Crabtree, B. (eds) (1999), *Environmental Indicators and Agricultural Policy* (Wallingford: CABI).

Manhoudt, A. (2006), Enhancing biodiversity on arable farms in the context of environmental certification schemes. PhD thesis, Leiden University, The Netherlands.

Melland, A.R., Love, S.M., Gourley, C.J.P., Smith, A.P. and Eckard, R.J. (2005), The importance of trust in the development and delivery of a decision support tool to reduce environmental nutrient losses from pasture systems, Published in the Proceedings of the International Congress on Modelling and Simulation 2005 <http://www.mssanz.org.au/modsim05/proceedings/papers/melland.pdf>.

Primdahl, J., Peco, B., Schramek, J., Andersen, E. and Onate, J.J. (2003), Environmental effects of agri-environmental schemes in Western Europe, *Journal of Environmental Management* Vol. 67, 129–138.

Schulze, P.C. (ed.) (1999), *Measures of Environmental Performance and Ecosystem Condition* (Washington D.C.: National Academy Press).

Smithers, J. and Furman, M. (2003), Environmental farm planning in Ontario: exploring participation and the endurance of change, *Land Use Policy* Vol. 20, 343–356.

8 *Farmer Learning in Benchmarking for Environmental Sustainability*

SUE KILPATRICK
University of Tasmania

The Implementing Best Practice in Sustainable Agriculture Project

Natural resource management (NRM) is a complex area. There is an imperfect scientific understanding of natural systems and how best to manage them. Furthermore, many players (including land owners and various government agencies) have overlapping responsibility for managing our natural resources.

The project described here worked with four groups of farmers in northern Tasmania, Australia over a period of between one and two years per group. Two project officers acted as facilitators, providing general agricultural knowledge input and identifying relevant specialists, experts and alternative management practices. The project funded the services of experts, as required. The project officers co-ordinated the activities of the group, initially intensively to assist group development, but with the aim of becoming less involved as the groups matured.

Each group first decided the environmental or NRM issue or problem they were going to monitor and benchmark. The process for developing and implementing best practice management in each group followed a similar direction but with different timescales for each step. The project officers guided the groups, drawing on a formal process for achieving continuous improvement and innovation that they developed. The NRM issues selected by the groups were nutrient run off and water quality (Group A), salinity (Groups B and C) and soil management for sustainable cropping (Groups B and D). Benchmarks were developed to monitor stream turbidity (Group A), salinity (Groups A and C), soil nutrient levels (Group A) and various attributes of soil health (Group B). Group D did not progress to develop any benchmarks.

Mapping the Groups' Activities

The activities of each of the groups at the various stages of the project were compared. The comparison was roughly chronological and identified the following stages: group initiation, selection of a NRM focus issue, identification of current practices and members' knowledge, selection of measurement tools and benchmarks, monitoring, action and continuous review. Aspects of the activities and characteristics of groups at each stage were compared (see Table 8.1).

There appeared to be a continuum of group effectiveness, ranging from one group who had developed performance and process benchmarks and taken action to achieve these (Group A) to Group D, which was arguably not performing as a group. In between, Group B had developed some performance benchmarks but not yet acted to achieve them, and Group C had commenced monitoring and considered possible benchmarks. Group A is the only group to have influenced others and the only group to have developed external networks. It was also the only group to have made or considered systemic environmental impacts (that is, those that extend beyond a farm boundary), albeit to a limited extent.

Two questions emerge:

1. How do the four groups compare to other measures of effective NRM groups? In other words, is Group A a highly effective NRM group compared to other NRM groups in Australia and internationally?
2. What are the characteristics of the groups that make them more or less effective and what has occurred in the groups (either before or during this project) to make them more or less effective?

Table 8.1   Aspects of groups at various project stages

Project stage	Aspects of group activities and characteristics
Initiation of group	Process of group selection
	Motivation of members
	Expectations of members
	Make-up and size of group
	Links between members at start
	First meeting
	Location of meetings and field/venue
Focus issue selection	Project officer meetings with individuals
	Facilitated structured 'focus session' with expert facilitator early on
	Type of early activities that helped build group in early stage
	Participation in training activities
	Selection of focus issue
	Influence of other current projects in selection of focus issue
	Member roles in establishing focus issue

Table 8.1   *Concluded*

Project stage	Aspects of group activities and characteristics
Identification of current practices and knowledge of group members	Identification of current practices and knowledge of group members
Selection of measurement tools and benchmarks	Role of other projects and other institutions/resources in gathering baseline data
	Role of members and others in selecting tools and benchmarks
	Nature of benchmarks
Monitoring	Monitoring tools
	How is the sampling done?
	Who does the interpretation?
	Group's ownership of monitoring process
Action	Planned actions
	Actions influenced by project
Evaluation and continuous review	Planned ongoing monitoring
	Influence of group on others
	Planned future group activities
	Evaluation

How Mature Are the Groups?

Pretty and Ward (2001) and Pretty and Frank (2000) derived four stages in the evolution of sustainable NRM groups:

Stage 0: Individualistic, use technology-derived solutions (*modernist system*).

Stage 1: Early group formation, either in response to a perceived crisis or prompted by an external agency. Outcomes tend to be adoption of practices similar to modern ones, but with less negative environmental impacts. Examples are low-dose pesticides and zero tillage (*reactive-eco-efficient-dependent system*).

Stage 2: Trust grows within the group, and rules, norms and links with other groups develop. Group members see they have the capacity to develop their own solutions and experiment. New practices tend to conserve and improve soils and water (*realisation-regenerative-independent system*).

Stage 3: Group members have acquired new 'worldviews' and ways of thinking, while groups maintain external networks, have a vision and are dynamic and productive. Groups are capable of influencing other groups. Agricultural systems are likely to be redesigned according to ecological principles and there are substantial improvements in performance or outputs (*active-redesign-interdependent system*). This stage involves such a ratchet shift for groups that they are very unlikely to unravel or, if they do, individuals will have acquired new worldviews and ways of thinking that will not revert.

Pretty and Ward (2001) list 15 indicators related to the stages of group evolution. The indicators are grouped into worldviews and sense making, internal norms and trust, external links and networks, technologies and improvements, and group lifespan. The typology represented by these stages suggests important relationships between group maturity and social capital, where social capital is networks and values or norms that enable people to work together for mutual benefit. They provide a useful frame for analysing the level of maturity of the four groups in this project.

Data gathered from the groups, the project officers, the steering committee, and technical and facilitator experts who interacted with the groups were analysed using Pretty and Ward's indicators, so as to compare the characteristics displayed by the different groups.

- Group A is the most mature group, displaying many characteristics of Pretty and Ward's most mature Stage 3, active-redesign-interdependence, particularly in relation to worldviews and sense making, internal norms and trust, and group lifespan criteria, but still exhibiting some characteristics of Stage 2, realisation-regenerative-independence, for example, in relation to technologies and improvements. It is at Stage 3 that groups are able to make systemic environmental impacts.
- Group B has made considerable progress in the 18 months since its inception, displaying many characteristics of Stage 2, but still some of Stage 1, reactive-eco-efficient-dependence, particularly in relation to the external links and networks criteria.
- Group C has just started the journey toward group maturity and effectiveness, displaying many characteristics of Stage 1, and some of Stage 2, for example a realisation of new capacities (one of the worldviews and sense making criteria) and sharing within the group (an internal norms and trust criterion).
- Group D was the most recently established group. It does have a small number of Stage 1 characteristics but has many more of the characteristics of Pretty and Frank's (2000) Stage 0. Stage 0 is a modernist system where there is no value put on social cohesion, individuals tend not to be organised in groups, and there is not effective monitoring to value natural capital adequately.

Pretty and Frank (2000) make the point that social and human capitals are prerequisites for natural capital improvements where the NRM issues cross property boundaries. This suggests that our question 'what are the characteristics of the groups that make them more or less effective, and what has occurred in the groups (either before or during this project) to make them more or less effective?' is particularly significant if we want to foster the development of effective, mature NRM groups.

Group Characteristics and Activities Affecting their Effectiveness

GROUP CHARACTERISTICS

The members of Group A, which had progressed furthest on Pretty and Ward's (2001) group maturity continuum, had worked together as a group on other projects and been involved in a number of community groups together. They socialised together and shared a range of values not restricted to those related to sustainable agriculture. Kilpatrick and Bell (2001) found that shared values relevant to the purpose of the group were a prerequisite for effective groups, and assisted in developing a shared vision of where the group is heading. Group A had a shared vision for NRM and sustainable agriculture in its district. Group B spent some time coming to realise that the members had shared values about sustainable agriculture. This group knew each other socially before the start of this project, but had not worked together before. Group C, which had not progressed as far on the Pretty and Ward (2001) continuum, also knew each other before the project, but did not interact to the same extent as Group B. Not all Group D members had met before the project. This continuum of 'baseline' member interaction and degree of shared values suggests that time and activities specifically designed to foster sharing and clarifying of values about NRM and sustainable agriculture are necessary if new groups are to develop into mature NRM groups.

There is some evidence from this project that the size of the group is related to effectiveness, with smaller groups being more effective. For example, Group A was the smallest of the groups, with six members, while Group C had 15 members. The important shared values and trust are easier to develop in a smaller group.

Group Process and Activities

SIMULTANEOUS GROUP AND TECHNICAL PROCESSES

The groups in this project have not only worked toward improving natural resources, or natural capital; they have also developed human and social capitals, with the possible exception of Group D. Further, given that the development of human and social capital is a prerequisite for improving natural capital, the group processes that have developed human and social capital are key to our understanding of how government agencies, communities and industry can act to facilitate the implementation of sustainable NRM practices.

The Implementation of the Best Practice for Sustainable Agriculture project proceeded through a number of sequential stages, from group initiation through selection of monitoring tools to action and continuous review. Kilpatrick and Bell (2001) described four sequential stages that groups go through as they develop and build social capital:

1. Acquisition of high levels of personal self-confidence and interpersonal skills, including leadership skills, is facilitated by building each other's self-confidence and by identity shifts (assisting members to see themselves in a new light) and building shared communication.

2. Getting to 'know' each other as individuals (including each other's history and future aspirations) and developing shared values and trust is facilitated by building internal networks: ensuring that members know each other's relevant knowledge and skills, and that there are shared values relevant for the purpose in hand (for example, about sustainable management of some aspect of natural resources).
3. Having group members come to regard each other as credible sources of support and advice is facilitated by building shared experiences.
4. Commitment to fellow group members, or being prepared to help each other out, is facilitated by building shared visions, or systematically reconciling past shared experiences with desired future scenarios (for example, ensuring that farmer knowledge and experience is recognised, valued and linked to future visions for the resources to be managed sustainably).

Analysis of the data relating to the four groups shows that the project officers and the groups are going though two simultaneous, sequential processes. One builds technical competency in NRM and the other is the underpinning social process that allows the groups to make decisions and work collectively. Attention to the detail of both the technical and group process stages is essential. The stages of the simultaneous technical and social processes do not completely overlap. Some of the activities and project officer actions with the four groups particularly assisted with group development and building social and human capital. They serve as examples of how the social and technical project processes can be linked.

The data from the four groups suggest that a preliminary stage should be added to Kilpatrick and Bell's sequential stages of group development. It is *foundation for building group relationships*, and is a set of actions that can establish a good foundation on which to build the technical and social processes. Examples of such actions are getting a farmer to champion groups, or the project officer arranging initial meetings and individual visits in early stages to get to know members' values and experiences.

At the time the focus issue was selected, the groups were building self-confidence and interpersonal skills. Activities that assist include practical activities, like soil sampling, that people are comfortable with, and a group training activity that gets people to interact. Groups are developing shared values and trust as they identify impacts of current practices, select monitoring tools and commence monitoring. Activities that help with both the technical and social processes are sharing information from the project officer's individually administered survey, the project officer consulting with experts and bringing information to the group to discuss, and looking for opportunities to build leadership in the group.

By the time benchmarks have been developed, best practice has been agreed and group members start to implement best practice, effective NRM members see each other as credible sources of advice. The project officer should assist in interpreting comparative baseline data in non-technical language and in such a way as to emphasise practices that may have led to the baseline measurements rather than just the figures themselves. The group should be prepared for individual members to adapt some practices to their own circumstances. There was little activity to observe in the project at the evaluation and review stage, as only Group A reached this stage of the project.

LEADERSHIP ROLES IN GROUPS

The role of the project officer changes as the groups develop. This is seen most clearly in Group A, and there is some evidence of a change in Group B. In the early stages of group development, the project officer acts as group leader. This is still the case with Group C at the end of the project. Some members of Group B expressed an interest in keeping the group together by initiating meetings and other actions. Group A is self-directed, coming together and making decisions on a range of issues not necessarily directed by the project officer. There is still a role for the project officer with Group A. It is one of co-ordination, liaison with external agents such as agencies, particularly those involved in monitoring, and as a channel to technical expertise. This co-ordination role is under the direction of the group: it is an 'executive officer' role.

The leadership in an effective group transfers from the project officer to group members as the group matures. The actions of the project officer in enabling and empowering group members to be leaders are crucial to achieving group maturity and so reaching Pretty and Ward's Stage 3 (active-redesign-interdependent). These enabling and empowering actions occur right through the social group development process. The seeds are sown by actions such as designing facilitation to give everybody a voice, which helps develop confidence, interpersonal skills and the group getting to know each other.

Effective groups share the leadership tasks. Again Group A serves as an example. Three leaders, each with a specific, well-understood role were identified from the six members. One person was an 'initiator', another was a 'driver', and the third keeps everyone on task.

Conclusion

A major factor in the effectiveness of the groups was the extent to which they developed shared values and visions in relation to sustainable agriculture. Having previously worked together and being part of the same social network assisted in developing shared values and visions. The activities of the groups influenced their rate of development as effective NRM groups. Activities that were matched to the stages of group maturity as well as the technical project stages were the most successful. Having the group attend a training course early in the project is an important example of an activity that helped members to get to know each other and develop shared values as well as acquire relevant technical knowledge and skills. Members taking on responsibilities within the group, including leadership roles, is important to group development and so effectiveness.

The groups' activities were determined by the project officer, except in the most mature group, where decisions were made jointly with group members. Hence, the project officer was a major determinant of group development and so effectiveness. The dual development of technical and social processes identified in this project mean that a project officer with generalist agriculture and NRM knowledge plus good group facilitation skills is required. The project officer should look for opportunities to give group members leadership roles.

The Implementing Best Practice in Sustainable Agriculture project benefited from drawing on a group action learning model. However, the most significant lesson from this project is that groups are going though two simultaneous processes: one builds technical competency in NRM and the other is the underpinning social process that allows the groups to make decisions and work collectively. Group activities must be designed to take both processes into account.

References

Kilpatrick, S. and Bell, R. (2001), Support networks and trust: how social capital facilitates economic outcomes for small businesses, in I. Falk (ed.) *Learning to Manage Change: Developing Regional Communities for a Local-Global Millennium* (Adelaide: NCVER), 79–87.

Pretty, J. and Frank, B. (2000), Participation and social capital formation in natural resource management: achievements and lessons, *International Landcare Conference, Melbourne* <http://www.affa.gov.au/agfor/landcare/pub/dof1-eval/references.html>.

Pretty, J. and Ward, H. (2001), Social capital and the environment, *World Development* Vol. 29, No. 2, 209–227.

PART 4

Benchmarking for Best Practice: Innovative Practice and Future Developments in Food and Farming

9 *Process Benchmarking and Other Developments in Agriculture*

Introduction

The previous chapters have set out the current state of play in benchmarking in farming and food. There are numerous mass participation benchmark systems in developed countries, and small group benchmarking practices established in New Zealand, Australia, the UK and more recently Canada. Environmental benchmarking is more difficult to locate – there are standards, indices and targets but the question is one of process: are they actually used in benchmarking for best practice? There is little evidence of this happening on a consistent basis, despite the number of well-developed and resourced initiatives in existence. In the food supply chain, management accounting practices are less advanced than might be supposed (Luther and Abdel-Kader, 2006), but there are very sophisticated data capture systems in the form of EPOS (electronic point of sale), ECR (efficient customer response) and RFID (radio frequency identification) that could feed into benchmarking for best practice systems.

The review of practices available reveals that there is potential for much more developed and sophisticated uses of benchmarking for best practice in farming and food, even given the highly developed nature of comparative analysis and discussion groups in the agricultural sector. In this chapter of the book, we explore individual businesses or groups that are using internal benchmarking and process benchmarking. The way ahead for sustainable change in agriculture is for incremental and continuous changes through the management of individual processes, activities and elements in farming. We argue that this will drive innovation in practice in the supply chain, on the farm and in environmental sustainability. The chapter is divided into four sections: a summary of findings in Chapters 2–8; innovations in internal benchmarking; innovations in process benchmarking; and challenges for the food supply chain.

Summary of Findings So Far

MASS PARTICIPATION BENCHMARKING

The use of benchmarks in agriculture for the purpose of comparative analysis is a long-standing practice that goes back to the late nineteenth century. There is a lack of empirical evidence to show how exactly farmers use the benchmarks on individual farms to make changes in their business plans and operations. We do know that many hundreds of farmers give data to various mass participation benchmark schemes and consider the performance of their farm relative to others. For some this may be simply a process of reassurance that they are 'not the worst', but results filtered by region or type of farm are useful for diagnosing whether a problem encountered or a downturn is systemic, local or national. Because the results are by definition anonymous, it is difficult to trace the reactions of individual farmers, except through the anecdotal or generalized experiences of advisors and consultants.

Mass participation benchmark data is used for policy making, advisory work, research and for broader interest: properly used, the data sets are a powerful diagnostic tool but they are, by definition, historic diagnoses. They are also a powerful research tool to enable monitoring of trends and the impact of interventions in farming. The information derived from such data provides starting points and end points, but in terms of the methodologies of benchmarking, there are additional approaches to best practice benchmarking that can be used towards achieving sustainable changes in the industry.

BUSINESS IMPROVEMENT GROUPS, OR BENCHMARKING FOR BETTER PRACTICE

In terms of sophistication, agriculture is ahead of businesses in other industries which have just realized the benefits of collaborative network learning groups. The small benchmarking group in agriculture has been established in New Zealand since 1949, and other countries have followed this model. The strength of the model is that it combines a high level of opportunity to share information – financial and physical – about farm businesses with a high level of socialization that is missing from many aspects of farming. By combining differing levels of information sharing, farmers do make strategic, operational and tactical changes to their businesses which have made those businesses in turn more sustainable. Experiments in Australia to harness the momentum of such groups to tackle environmental problems are a step towards using the model to address individual problems or processes. Evidence from those working with improvement groups, including a technical director in a major estate agency in England, suggests that as a group progresses, the members identify areas of significant concern or amenability to changes and focus on individual items – labour and machinery for example. A case study from the Food Chain Centre (FCC) tells the story of one group that focused solely on veterinary costs. This indicates that business improvement groups can evolve into what is essentially process benchmarking, a theme which is developed later in this chapter.

The weaknesses of such groups are practical. The dynamics of the group and the enthusiasm and experience of the group facilitator are significant both to the success of the group and plans developed on the findings of the group. Many groups are based geographically, which increases the 'like for like' measurements in the group but also decreases the chances of measuring against the 'best in class', wherever that is. Hence the title of Part 2: benchmarking for *better* practice rather than for *best* practice. The potential for incorporating into the model elements of what elsewhere are called competitive benchmarking and best practice benchmarking are explored below.

ENVIRONMENTAL BENCHMARKING

Environmental benchmarks and best practice initiatives are varied and widespread in the industry, but have not coalesced into similar patterns of practice in the same way that mass participation benchmarks and discussion groups have done. Benchmarks in environmental best practice tend to be presented in the form of standards, certification requirements or codes of practice. In many cases, participation or take-up of schemes is driven either by legislation (in the case of nutrient levels) or subsidy (for European and American agri-environmental schemes), or by incentives offered by supermarkets for environment-friendly produce. The exception appears to be Australia, where extensive engagement with Landcare groups is driven by chronic environmental problems that threaten the sustainability of agriculture in the country. There are benchmarks and a search for best practice, but not as yet benchmarking for best practice in a coherent form. The aim is surely to incorporate the search for processes that are environmentally sound into discussion groups and process benchmarking as they develop in the sector.

Moves Towards Process Benchmarking

It is clear that benchmarking practices have become embedded into farm management and that there are many innovative approaches to achieving better practice both in benchmarking and in adapting best practices of others to create businesses that are more viable and more sustainable. Yet it is also clear that benchmarking for best practice (as understood in 'conventional' accounting for manufacturing and service industries) happens only in a small number of individual cases. What is missing is a clear methodology and drive towards using *process benchmarking* in agriculture and food on a wider scale, to complement existing methods of benchmarking. Those businesses which are too large or too dissimilar to other farm businesses to meaningfully participate in comparative analysis databases or discussion groups without skewing the results, or being the best practice model for others without achieving changes themselves, should be examining the use of best practice benchmarking on a one-to-one basis. One farmer said, 'We don't do benchmarking because we're not like anyone else.' Yet there is scope to look at operations overseas or in other industries. On a smaller scale, it is apparent from discussions with group facilitators and from some published case studies (www. foodchaincentre.co.uk) that discussion groups isolate and focus on individual problem areas – for example, machinery and labour, or vets bills – and begin in-depth analyses of the processes underlying the numbers. The beginnings of process benchmarking in agri-food can be seen and are the next development for consultants and other advisors to facilitate in the industry.

The Elements of Process Benchmarking

When *Benchmarking: An International Journal* or other practitioner-led resources in management consultancy and accounting talk about benchmarking for best practice they are primarily referring to process benchmarking, which has been developed from the idea of one company approaching another to share data about a single process in the business in order to achieve efficiency and economic gains through improvement. There is adaptive innovation in taking the ideas of one company and reinventing them in another setting. Since the Xerox Corporation successfully improved its warehousing operations by comparing and mapping them against those of L.L. Bean, even though that company was from a different sector, benchmarking for best practice has been developed into a major strategic management tool.

A number of writers have already called for process benchmarking to be developed and adapted for the agricultural industry. The most notable article comes from Australia in 2000 – so the calls are, in terms of diffusion of management ideas, still at an early stage.

Ronan and Cleary (2000) 'agree with critics that much of what is currently called "benchmarking" is difficult to distinguish from comparative analysis, lacks systemic linkage to underlying enterprise processes and drivers of competitiveness and is of limited diagnostic power at farm, supply chain and industry levels'. In their opinion, benchmarking practice needs to get away from comparisons of gross margins and comparative farm profits – the financial measures – and into analysis of processes and procedures which largely involve non-financial measures. The first step is to clarify what is meant by processes and by process benchmarking.

Generic Process Benchmarking

The organization processdriven.org identifies four generic types of process benchmarking:

1. *Internal benchmarks* that look inside an organization to see if there are improvements to be observed from other locations and groups.
2. *Competitive (or external) benchmarks* that look outside the organization to see what can be learned and observed from processes at other companies in the same industry.
3. *World-class benchmarks,* which extend the external observation to companies in other industries that are considered to be the best in the world at what they do.
4. *Activity-type benchmarks* that focus on a narrow part or piece of a process, again transcending industries in the search for new approaches or solutions.

The key to process benchmarking is that it is not primarily a comparison of financial results and indicators, although these might be used, but an analysis of processes, mapping out each step (say of an invoicing and debt collection procedure, or of a production line in manufacturing concerns), identifying how things can be done better and re-engineering the process to achieve a better process for one's own operations. A business *process* is a group of logically related tasks (serial and parallel activities) that takes inputs and transforms them to create output that delivers customer value.

One organization, BenchmarkingPlus, identifies the following characteristics of process benchmarking:

- *Focus*: a single process at a time.
- *Partners*: not chosen until after undertaking a thorough analysis of your own practices and performance.

- *Form of comparison*: whenever possible, by actually visiting the partners' places of business.
- *Confidentiality*: the identity of partners is known, and the exchange of information is protected by a code of ethics.

The method can be as shown in Figure 9.1.

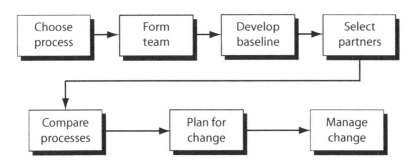

Process Benchmarking (simplified)

Figure 9.1 Schema of process benchmarking

© Benchmarking PLUS 1999.

Bendell et al. (1993) use a different terminology, which is also useful in identifying what is meant by process benchmarking. They, too, identify internal and competitive (external) process benchmarking as common approaches to finding improvements through the comparison of similar operations. Their other two categories are *functional benchmarking* and *generic benchmarking*. Functional benchmarking involves comparisons of similar functions within the same broad industry or sector through non-competing organizations that carry out the same functional activities (which could be warehousing or administrative functions, as well as areas of production). Generic benchmarking is the same type of comparison but carried out in industries which are different to that of the benchmarking organization, which can lead to innovative transfers of ideas between industries.

Ronan and Cleary (2000) identified the benefits of this approach to agriculture:

Benchmarking does some things that cannot be expected to gravitate out of individual business analysis or a general awareness of new technology and management innovation possibilities. Process-based benchmarking permits a sequence of data inquiry, mining beneath broad business and enterprise performance outcomes down to processes influenced by management and the environment. It identifies and illustrates production processes which, through systemic linkage to operational approaches, business structures and supply chain arrangements, if changed, can result in improved productivity and profitability. It focuses on the key drivers of competitiveness that managers [can] control.

Although a concerted follow up to Ronan and Cleary's (2000) challenge to consultants and others to take up process benchmarking is yet to be seen, a study in 2008 sets out a hypothetical model for using process benchmarking and non-financial indicators in the poultry industry. The authors, who have written elsewhere about benchmarking (or the relative lack of benchmarking) in the food industry, observe:

> *Benchmarking approaches have evolved from initial cost-focused comparative analysis to process-orientated benchmarking. Indeed, it has been argued that if not effectively implemented, livestock system benchmarking techniques can focus too much on historic data rather than identifying and implementing current best practice and facilitating knowledge transfer. Effective livestock supply chain benchmarking is more than a comparative analysis of cost structure. It requires a detailed understanding of the processes undertaken in order to determine the ideas and information that needs to be shared both vertically and horizontally in the chain which in turn will deliver compliance with stakeholder requirements and drive continuous improvement.*

(Manning et al. 2008: 161)

Research in this area has shown that there are individual cases of farm businesses using forms of process benchmarking. The first typology of process benchmarking given above identified the following classes of benchmarking:

- internal;
- external;
- activity;
- world-class.

Concerning the last of these classes, comparison of farming practices across countries and transfer of knowledge does happen in a number of ways:

- formal and informal farm visits in other countries;
- conferences through organizations such as the International Farm Management Association;
- collection of international comparative data such as the FADN in Europe (discussed in Chapter 2) or the International Farm Comparison Network, which produces a Dairy Report, for example, giving statistics and measures across some 40 countries, which feed into policy level discussions but which could also form a basis for identifying best practice farm businesses in other countries.

International benchmarking is an extension of external benchmarking, but there appears to be as yet no concept of world-class benchmarking in agriculture and food, and this has been regarded as outside the scope of this book.

The next section sets out short case studies for internal, external and activity benchmarking. Internal benchmarking is being used formally on corporate farms, and more informally by farmers operating farm contracting services who need to monitor the efficiency and costing of each contract they enter into. External benchmarking can be seen in a group that is formed solely to monitor machinery and labour costs of commercial machinery/labour rings. A growing form of activity benchmarking is in the area of veterinarian issues and disease management, with innovation being driven by veterinarian practices. Finally, the notion of using non-financial measures has been developed and tested at an academic level, and is being promoted through balanced scorecard projects in New Zealand, the US and the UK. Chapter 10 is an extended study of the work and potential of the balanced scorecard as a benchmarking and business improvement tool.

Innovations in Benchmarking for Food and Farming

INTERNAL BENCHMARKING

Internal benchmarking is most applicable in corporate farming organizations, where there are a number of farms or farm units engaged in similar enterprises but run under one business heading, or in contract farming, where one farm business may manage a number of others alongside its own. Such businesses are likely to run an accounting operation and collate financial and physical data on central databases. At the very least, it is likely that each farm unit will (or could) be running the same farm management and accounting software packages. What makes internal benchmarking work, however, is the presence of an individual – a 'champion' – who is prepared to collate, analyse and interpret the figures. Furthermore, this would lead to action through training and persuading farm managers throughout the group to look at how best practice on the better performing farms in the group could be replicated through the less well performing farms. Another aim is to look at practices outside the business and adapt innovations elsewhere to bring down costs and improve yields and margins.

Sentry Farms case study

Sentry Farms in the UK is the best known exponent of internal benchmarking and cited regularly in the agricultural press. A leading expert in the area referred to Sentry Farms as one of the two farming businesses in the UK who 'really did farm management properly'. The following extract from *Farmers Weekly* illustrates their approach, which fits very neatly into the category 'process benchmarking':

> *Fuel use across 18,279ha (45,167 acres) of Sentry-managed farms has declined by an average of 2% a year over the past 15 years or so, and the company now uses about 40% less fuel per hectare than it did in 1990.*

> *'Until recently, fuel has never really been seen as a significant cost on farms, as prices have been about 20p/litre. Therefore there's been very little work done recording fuel use. But, the situation has changed dramatically – last year our fuel spend doubled – and we, as farmers, have got to do everything we can to keep that 2% trend coming down.'*

> *'Substituting last season's Sentry benchmarking data with current diesel prices highlights how drastically things can change,' he says. 'Within two years profitability drops from 2007 levels by 74%. If you're using 100 litres/ha of fuel, for example, every time diesel price goes up 1p/ litre, that's £1/ha straight off the bottom line.*

'This makes it even more essential to accurately record fuel use at the farm and individual field and operation level, and get to grips with how much fuel is used to produce one tonne of wheat, or total fuel cost per ha or per hour.'

(P. Spackman, Tips for improving diesel efficiency on your farm, *Farmers Weekly*, 6 June 2008)

A detailed study of internal benchmarking by Sentry has already been published by the Food Chain Centre and is available on www.foodchaincentre.com (Business Clubs). The process of creating the internal benchmarks is an intensive one, carried out by the technical director over two to three months in the winter period and with the results being presented at an annual conference in February. In addition, the data can be used as comparative data for farms to which the group offers consultancy, as well as being used internally. The director champions the use of the data, with results similar to those reported above being achieved elsewhere in the business in other processes and activities. The data also informs decisions about target setting and drives the search for innovative practice. The following comments from the Food Chain Centre study in 2005 highlight the commitment of the company to its benchmarking practice.

Sentry began benchmarking soon after its formation in the 1970s. The broad range of farms within a single organisation gives Sentry a strong base for benchmarking. By banding together, they can employ experts, such as their technical director. He says: 'Benchmarking allows our farm managers to learn from the trials and tribulations of other units, without having to make the same mistakes themselves.'

Sentry don't believe in simple league tables showing who has the best farm. They know that different farms excel in different aspects. So everyone in the group learns from each other. They use a wheel to illustrate benchmarking scores. A score near the centre of the circle is low and near the rim is high. So the example farm in red has a higher cost than average in most categories.

(*Benchmarking through Business Clubs Case Study*, May 2005, Food Chain Centre)

Internal benchmarking elsewhere in the food chain

A *Financial Times* report in 1998 into best practice in management excellence in the food industry identified a few examples of best practice benchmarking. Later reports suggest that these still remain best practice in the industry. One of the frontrunners was Frito-Lay International, which had highly developed reporting systems both internally and with suppliers that contributed to improvements in practice. The internal system involved 'a 100-page [monthly] report that compares all its manufacturing sites' performance on quality and safety using the same performance measures. This report, distributed to all sites, encourages the weaker performers to contact the best sites in order to share ideas and improve' (Leatherhead Food RA, 1998: 83). The success of this report, the format of which was developed with the input of those receiving the information, had encouraged them to roll it out to suppliers. Frito-Lay makes supplier relations a priority (highlighting on its web pages its use of women and minority suppliers) and incorporates suppliers into its internal performance measure and benchmarking systems. The *FT* report comments that 'farmers are requesting real-time production data [from Frito-Lay International] to improve their understanding of the relationships between the quality of the crisp and the quality of the potato' (ibid.) Furthermore, the company was holding at that time regular meetings with its suppliers (in groups of six to seven) to design a monitoring system for product quality and safety, the aim being to facilitate contacts between suppliers to carry out what is essentially process benchmarking. 'The suppliers see benefits from working with other suppliers as they do not see them as competitors' – in order to protect the company and its trade names (Walkers Crisps in the UK) they work together to protect the supply base. Frito-Lay had also formed close partnerships with a small number of packaging suppliers, following policies of its parent company, PepsiCo. (ibid. 39).

The sharing of information for benchmarking by suppliers of a particular organization is growing. Veterinarian groups are discussed below, but another link in the food supply chain where these practices are being attempted is abattoirs. A story in the UK farming press reported, 'Beef farmers supplying [an Aberdeen-based] abattoir are to receive regular feedback bulletins, allowing them to benchmark their performance against the plant's average. Producers' summaries will report on carcass weights, conformation and fatness scores, as well as estimating liveweight gain compared to the factory average … Benchmarking summaries will also indicate values for different carcasses, allowing producers to see differences in returns' (*Farmers Weekly*, 28 October 2004). The Agrifood and Biosciences Institute (a British governmental organization) offers the Pig Grading Information System (PiGIS) developed in Northern Ireland as an online system for the sharing of benchmarking information in the area of pig carcass quality.

Details of benchmarking practices in food processing and retail firms are not generally publicly available and there are no detailed case studies to draw on. The evidence is that the practice is not widespread but has emerged and is growing. Another example in the press is that of Kellogg's:

> For benchmarking purposes, all Kellogg factories use the same monitoring techniques, so it is possible to compare performance between sites, although there are always some things that are done differently. We can interrogate this information to improve performance across every site. As things have improved, we have also had to reassess our baseline figures and develop more sophisticated tools to monitor performance to ensure we continue to make progress. We've seen a 20% increase in productivity in six years using this system.
>
> (Federic Roquet-Jalmar, Benchmark for success, *Food Manufacture*, 2 October 2007, www.foodmanufacture.co.uk/news/fullstory.php/aid/5379/Benchmark_for_success. html.)

Internal benchmarking between corporate clients and supplier groups based on non-financial information and process mapping is growing, but the potential for exploiting and generating best practice information from the sector is as yet largely untried.

EXTERNAL BENCHMARKING

External or competitive benchmarking in agriculture as found in other industries is not apparent; however, the following case study involves a group that are sharing information, but as competitive profit making businesses rather than as the more usual co-operative arrangement. The more arm's length nature of this group suggests that it is more in the area of external process benchmarking, although its approach could be described as activity-based benchmarking as well.

Joint Venture Farming Group case study

The Joint Venture Farming Group (JVFG) in East Anglia, UK comprises a number of joint ventures between farms, taking the form of 'machinery rings' and/or 'labour rings'. The group's objectives, stated on its website (www.jvfg.co.uk) are to:

- provide a forum for owners, managers and employees involved with joint venture farming to exchange knowledge and ideas;
- provide a forum for benchmarking physical and financial performance;
- provide a forum for political lobbying (via the National Farmers' Union) on matters that affect joint venture farming businesses;
- provide a force to influence machinery manufacturers;
- provide a means of communication between members;
- provide training for directors and staff;
- explore any potential commercial opportunities for the benefit of the group and its members.

Machinery and labour rings in the UK exist in a number of forms, with a co-operative structure being the most common. The Borders Machinery Ring claims to have been the first to be formed, in February 1987, and states its aim as 'rationalising labour, machinery and input costs' (www.ringleader.co.uk/). Other rings have grown to include purchasing and labour agency work. The majority are not-for-profit organizations. The JVFG rings differ in that they are essentially run as separate enterprises or profit centres by the businesses concerned and there is a legal profit sharing agreement between businesses. It is possible that this more commercial aspect to the formation of the machinery and labour rings has driven the move to benchmarking performance.

The general concept of a machinery ring is to put 'suppliers' who can provide equipment and labour in touch with those needing work done, 'demanders', with a manager or co-ordinator to organize rotas whereby members of the ring have the use of machinery on an equitable and timely basis. Grant Thornton set out the benefits of ring membership as:

For Suppliers – access to a larger market place, there for potential for increased amount of contracting work and reduction in the need for advertising. Administration looked after by the ring manager. Prices set by the ring and negotiations with customer not required. Reduced risk of late or non payment. Utilisation of machinery and labour outside peak seasons. May justify larger machinery and therefore increase timeliness of operations on suppliers own farm, or justify ownership of specialist machinery which may be in demand.

For Demanders – ring manager matches requirements with services available, often planning several months in advance, and therefore a more reliable service is guaranteed, with likelihood of contractor arriving late being reduced. Access to larger or more specialised machinery. Prices set. Only need to deal with one person – the ring manager. Minimum hire periods can be utilised – say 12 weeks, can be used by several members therefore justifying cost. Timeliness and continuity guaranteed with back up or replacement from other members.

(Markham and Chapman, 1998)

The members of the JVFG run as private limited companies rather than co-operative arrangements. Each company has its own set-up, but essentially the JVFG provides machinery and labour its members, and to some third-party businesses, and aims to make a profit doing so.

The idea for collaborating on a national level, sharing financial and physical data and ideas, for benchmarking and business improvements was mooted by one of the current members of the group in 2004. Sponsorship and support was obtained from manufacturer CLAAS/Simba, the English Food and Farming Partnership, the National Farmers' Union (NFU), Grant Thornton and the East Midlands Development Agency. Members also pay an annual fee for the service. This enabled a software developer to be commissioned to build the benchmarking programme. As well as the data processing, the group has an annual conference, farm visits, training sessions and discussion group meetings.

Benchmarking data is constructed from audited financial and physical data, either input directly or delivered to the system manager and input centrally. The aim is to analyse cost structures and to identify best practices that will reduce costs and increase efficiencies across the group. Results can be filtered to be across the farms in a member company as well as comparative across the whole group. There are a variety of reports, including results for combinations of machinery, machine idle reports, coverage and global financial results, such as that in Table 9.1.

Table 9.1   Joint Venture Farming Group results for three years on key indicators

Operations costs	2005 Harvest £/ha	2006 Harvest £/ha	2007 Harvest £/ha
Primary cultivations	18.75	32.85	27.92
Secondary cultivations	8.41	19.43	16.34
Drilling	12.71	19.53	19.32
Rolling	4.40	5.63	5.82
Fertilizer application	5.43	3.65	3.96
Spraying	3.54	3.34	3.18
Harvesting	49.51	48.58	46.84
Carting	14.66	12.65	15.44

Results are then analysed further to look at:

- crop establishment costs;
- harvest costs;
- fuel usage;
- fuel efficiency;
- costs per hour and per hectare per operation;

Discussions then continue to examine questions of purchase versus hire of machinery, cultivation techniques that can reduce fuel usage, best combinations of machinery, effects of low-tillage systems and ways of minimizing break-downs. It was noticed that there was a difference of 17 per cent in operating costs across the group for wheat and 34 per cent for oilseed rape (canola), and thus scope for improvement of practice. In the future, the group hopes to be able to develop interfaces with crop, accounting and wage records, and compare charge out, depreciation rates and standard unit rates. Innovative solutions to data input (such as capturing timesheet data on handheld devices) are being investigated, and to presenting the data in easily digested formats (such as traffic lighting).

The JVFG combines both best practice in collating data and running collaborative small group network benchmarking that characterizes agricultural benchmarking practices, but it is also an example of how process benchmarking is beginning to work in farming businesses. One limitation on the depth of analysis possible is the time that farmers are willing to commit to providing and analysing data, or the cost they are willing to pay for this to be done for them. The group is heavily dependent on sponsorship as investment, though obvious economies of scale are achieved by running the benchmarking system as for a group rather than for one corporate company. The strengths of the system are in being able to identify improvements and to bring down costs across the group through identification of best practice. As with any such venture, the future development is dependent on the commitment of the businesses involved.

ACTIVITY BENCHMARKING FOR ANIMAL WELFARE

Defining activities in farming or processes has not been attempted and in fact could differ from farm to farm business. One clear area that has been identified is livestock herd health. The drive for this has come through veterinarians and animal health researchers. The key outcome has been the use of non-financial indicators to score lameness and to compare practices on different farms to reduce incidence of diseases such as mastitis. This is known as condition scoring. The following article indicates the nature of condition scoring used on a British dairy farm:

> Dry and fresh-calved cows were routinely condition scored to track overt changes. Blood samples are also now taken monthly from 10 freshly-calved cows. These are analysed for beta-hydroxybutyrate (BHB), an indicator of energy balance, which can give warning of impending ketosis.

> 'The individual cow BHB results are categorised using a traffic light system – red for danger, amber as a warning and green for OK, and plotted on a graph. Any trend towards more cows

being in the red than green are soon highlighted. Diet alterations can then be made to provide more energy,' says the vet.

Clinical mastitis had been a problem in the past due to chronically infected cows and contagious mastitis pathogens. But the farmers and the vet had already reduced this through focusing on parlour hygiene and culling problem cows.

Benchmarking against other herds within the Millcroft vet practice, showed the Paddle herd was just above the median figure for 2007. So the help of a milking technologist was enlisted. Ventilation in cattle buildings can influence mastitis incidence, says the technologist. 'Poor airflow leads to humid buildings and in moist, warm conditions bacteria can thrive and environmental mastitis can become a problem.'

(Lawrence, *Farmers Weekly*, 19 March 2008)

One particular form of condition scoring is for lameness (and so referred to as *lameness scoring*). Amory et al. (2008) state that lameness in cattle 'is of huge economic importance with the current high prevalence of lameness in dairy cows, estimated to be 15 per cent in the USA ... and 22 per cent in England. Milk loss per cow because of lameness has been estimated to be 440 and 270 kg for early lactation and mid to late-lactation, respectively, in France ... and up to 2 kg/day for up to 5 months before and after diagnosis in the UK' (381).

DEFRA in the UK have issued a report (produced by ADAS) in which they summarize the issue of lameness by saying:

It is painful to the animal, it is a serious welfare issue and is costly to the dairy farm business. The cost of lameness is estimated ... to be in the region of £178/affected cow...

The main husbandry factors involved with dairy cattle lameness are:

- *Environment (housing and underfoot conditions);*
- *Nutrition;*
- *Breeding;*
- *Cow behaviour;*
- *Youngstock management.*

(DEFRA/ADAS, 2007)

Use of lameness scoring tools is geared to identifying individual herd management processes that can then be assessed and re-engineered where necessary. It also opens up the potential for useful benchmarking activities between farms to manage the risks and improve practices to reduce the distress and cost associated with animal disease. Amory et al. (2008: 390) summarize the approach succinctly by saying that their research 'emphasises the importance of recording the lesion-specific causes of lameness to determine both the possible economic consequences for a herd and to inform on management decisions to reduce lameness'. A veterinary advisor to the UK supermarket chain Tesco in their farming initiatives states that mobility scoring, as a management tool to assist farmers in identifying and prioritizing cows for treatment, is a key aspect of managing a herd's health.'Mobility scoring identifies cows in the herd that require additional care and helps producers identify whether the management of their herd is effective at reducing the level and effect of lame cows. These management measures include foot trimming schedules and foot bathing protocols in addition to looking at the comfort of cow housing on farm and the management of calving cows within the herd.' (www.tescofarming.com/v2/mobility.asp).

An example of lameness scoring methods is found on the Bristol University Vet School Healthy Feet Project website (www.cattle-lameness.org.uk/index.php).

In the poultry industry, the Manning et al. (2008) paper referred to earlier presents a hypothetical model for using process benchmarking. They identify traditional performance measures as being based on mortality or disease rates and feed conversion rates, with cost-based measures such as financial returns per bird, financial returns per kilogramme liveweight or per square metre per week as being inadequate to effect sustainable changes in practice in poultry meat production. Their research found that more important non-financial factors concerned water usage, food quality, lighting and heating, the underlying manageable risky factors to ability to convert feed to meat, disease control and mortality. Improve these areas and compare these processes across farms, and the potential for meaningful benchmarking for best practice emerges. What is required in the industry is more research of this nature, identifying the non-financial benchmarks and the activities that underpin operations and give the best chance for benchmarking to contribute to sustainable change in farming and food practices.

Other livestock practitioners have identified water and energy usage as key factors in managing and improving operations, along with feeding patterns. A number of benchmark indicators and standards are becoming available, through partnerships with retailers, such as Tesco's dairy standards, which members of the Tesco Sustainable Dairy Project launched in 2007 are required to adhere to, or the benchmark figures produced by the UK National Pig Association in relation to energy use in pig farming.

The growth of research and development of newer practices and performance measures in this area is increasing the scope of benchmarking in agri-food into a process benchmarking for best practice and away from comparative analysis as the sole basis of farm management improvements.

BALANCED SCORECARDS

Process benchmarking, as seen above in the lameness in cattle project, is as much or even more concerned with non-financial indicators than financial ones. A recent development in performance measurement in agriculture is to try to establish the use of balanced scorecards (BSC) based on the Kaplan and Norton model (see Chapter 10). This model has four perspectives: the financial perspective, the customer view, the internal operating perspective, and the innovation and learning perspective. The work in this area has been pioneered by Massey University in New Zealand, and this project is discussed in Chapter 10 by Nicola Shadbolt. There have been two other notable BSC projects dating from around 2006, led by the Food Chain Centre and Writtle College in the UK (www.foodchaincentre.com) and by the South Dakota State University in the US (Dunn et al., 2006).

When identifying which measures should be used in best practice benchmarking, Bogan and English (1994: 65) comment, 'the balanced scorecard offers a structured framework for thinking about performance measurement and for constructing integrated performance metrics ... [However] the scorecard elements are single data points that together describe an operating system but do not provide comparative information that establishes the adequacy of individual performance levels.'

The South Dakota/Texas ranch model of the balanced scorecard uses six perspectives, to recognize particular perspectives that are more peculiar to farming (Dunn et al., 2006):

- learning and growth;
- natural resources;
- agricultural commodities/production;
- customers;
- financial;
- ranch lifestyle.

However, in terms of benchmarking, the balanced scorecard is seen as a user rather than a producer of benchmarks. The performance indicators generated by management should be derived from, or 'reality checked' against, other benchmarks. Dunn et al. (2006) suggest ranch managers should use 'measurements that can be compared to a benchmark that has been created using the same terms, definitions, and protocols. The benchmarks should be from relevant geographical areas and be up to date. Appropriate key performance indicators in the agricultural commodities/production perspective for a cow-calf operation might include pounds weaned per female exposed, pregnancy rate, replacement rate, cow body condition score at weaning' and so on (ibid. 18).

Developing Robust Process Benchmarking for Sustainable Change

In agriculture, the call for best practice benchmarking to be developed and for the industry to move beyond comparative analysis of financial indicators alone has been best articulated by Ronan and Cleary in Australia:

> *Best practice benchmarking systemically links processes and performance, provides a balance of production, financial, environmental and social indicators and presents information which enables easy, unambiguous interpretation by farmers. It is not a substitute for production economic analysis. It can aid the pursuit of productivity and profits by farmers through richer information about enterprise performance and can provide a catchment of ideas for continuous improvement.*

> (Ronan and Cleary, 2000)

In other industries competitive benchmarking is identifiable as the most common form of best practice benchmarking (Bendell et al., 1993). Since Xerox, best practice benchmarking has been refined and lessons have been learned that could be used to improve practice in farming and food. At the same time, the natural evolution, examined in this book, of network and learning benchmarking in which individual practices and processes are identified, isolated and subjected to scrutiny within groups could in turn inform practices in other industries. The main difference between best practice benchmarking in other industries and in agriculture is that in the latter, groups tend to be formed around existing networks: levy boards, consultancy and accounting firms, universities, corporate farming businesses and geographic location. In other industries, individual firms identify and approach firms worldwide that are felt to have best practice and form partnerships. This introduces a level of management, tact and diplomacy into the benchmarking process and it is this, as much as the techniques used to map and evaluate processes, that forms the basis of much of the writing on best practice benchmarking.

If best practice benchmarking is to evolve in farming and food, and to offer opportunities to those farms that have gone as far as they can with comparative benchmarks and discussion groups, then protocols for best practice benchmarking need to be developed. The groundwork for change in the culture of farming so that seeking information and becoming 'information rich' (Verissimo and Woodford, 2005) is now part of farming, has been accomplished over many centuries in agriculture. The diffusion of the practice of best practice benchmarking as acceptable on an individual business-to-business model has not yet been seen but would provide one more tool alongside the other forms of benchmarking in the industry to create sustainable change.

What is best practice in best practice benchmarking? Bogan and English (1994: 114) make a very important point, which is that 'benchmarking, like other management initiatives, is subject to learning curve effects'. Thus it tends to 'grow easier, faster, and more effective with practice and experience'. The key problem of inexperience and poor planning and discipline in benchmarking is that an inordinate amount of time is spent on data gathering. Bogan and English observe in the same discussion that:

> *Benchmark teams that achieve optimal results also invest in adequate time and resources in early planning and in the communication and buy-in process. They recognise that these phases may seem less glamorous than hopping on a plane to visit another company. Yet these phases can have a disproportionate influence on the project's ultimate success*

They identify a number of 'secrets of successful benchmarking':

- training;
- information technology;
- cultural encouragement of learning;
- resources;
- design of the benchmarking process;
- implementation and business re-engineering.

There is no standard process for process benchmarking and there is much to be said for customizing practices for each business or organization partnership taking part. In agriculture, there are already resources in place to promote benchmarking and, almost uniquely, there is a culture of learning and collaboration which many more competitive commercial environments would find difficult to begin to build. In order to establish process benchmarking as a complement to other forms of setting benchmarks and small group benchmarking – another tool available to farm businesses – consultants, academics and other industry advisors need to turn their attention to the following matters:

* identifying the processes and underlying drivers;
* identifying what is manageable by farmers and what is not;
* setting up a protocol and a facilitation process to carry out mapping of processes and analysis;
* imaginative search for benchmarking partners – abroad and outside the industry;
* moving away from financial accounting based systems of the whole farm and moving towards unit costs – but one that allows for proper returns to the farmer for reinvestment, reward for labour, knowledge and investment, and incorporates costs of environmental protection, animal welfare and other societal goods.

Sustainable Change through Benchmarking in Farming and Food

Successful benchmarking in farming and food comes through collaborative arrangements with peers or between customer and supplier which are based on trust and mutual benefit. In this regard, the industry is ahead of many others in which competitive, limited scope benchmarking has been developed. Yet there is potential for more individual benchmarking exercises to be carried out, which would in time be disseminated through the industry as best practice. In farming, there is extensive evidence that farm visits, demonstration and monitor farms, and other activities that make the operations of the farm visible are the key strength of benchmarking as they enable participants to visualize sustainable change. Lengthy participation of discussion or environmental groups for example, is linked to improved financial well-being and survival. The financial basis of mass participation benchmarking is evolving into process benchmarking as discussion group members concentrate on single pervasive issues and begin to drill down to see what in the processes behind the numbers is driving cost. The next logical step is to develop one-to-one benchmarking with partners further removed geographically or in other industries, to complement existing practices. The sustainable change is linked not to the techniques that are used but to the learning that is facilitated – and that is the lesson that could be carried throughout the food chain and into other industries.

References

Amory, J.R., Barker, Z.E., Wright, J.L., Mason, S.A., Blowey, R.W. and Green, L.E. (2008), Associations between sole ulcer, white line disease and digital dermatitis and the milk yield of 1824 dairy cows on 30 dairy cow farms in England and Wales from February 2003–November 2004, *Preventive Veterinary Medicine* Vol. 83, 381–391.

Bendell, T., Boulter, L. and Kelly, J. (1993), *Benchmarking for Competitive Advantage* (London: Pitman Publishing).

Bogan, C.E. and English, M.J. (1994), *Benchmarking for Best Practices: Winning Through Innovative Adaption* (New York: McGraw-Hill).

DEFRA/ADAS (2007), *Dairy Cattle Lameness – Practical Solutions to a Persistent Problem* (London: DEFRA) <http://www.defra.gov.uk/animalh/welfare/farmed/advice/cow_lameness.pdf>.

Dunn, B.H., Gates, R.N., Davis, J. and Arzeno, A. (2006), *Using the Balanced Scorecard for Ranch Planning and Management: Setting Strategy and Managing Performance*. (South Dakota State University/Texas A&M University-Kingsville) (EC922) <http://agbiopubs.sdstate.edu/articles/EC922.pdf>.

Lawrence, C. (2008), Farm health planning: routine vet visits are vital, *Farmers Weekly* 19 March 2008.

Leatherhead Food RA (1998), *Business Excellence in the Food and Drink Industry: Practical Ways to Improve Performance, FT Management Report* (London: FT Retail and Consumer Publishing).

Luther, R. and Abdel-Kader, M. (2006), Management accounting practices in the British food and drinks industry, *British Food Journal* Vol. 108 No. 5, 336–357.

Manning, L., Baines, R. and Chadd, S. (2008), Benchmarking the poultry meat supply chain, *Benchmarking: An International Journal* Vol. 15 No. 2, 148–165.

Markham, G. and Chapman, T. (1998), *Investing in Farm Machinery* (London: Grant Thornton).

Ronan, G. and Cleary, G. (2000), Best benchmarking practice in Australian agriculture: issues and challenges, *Agribusiness Perspectives*, Paper 39 <http://www.agrifood.info/Review/Perspectives/2000_Ronan/2000Ronan>.

Veríssimo, A. and Woodford, K. (2005), Top performing farmers are information rich: case studies of sheep and cattle farmers in the South Island of New Zealand. Published in the Proceedings of the Fifteenth International Farm Management Association Congress, Campinas, Brazil, August 2005, Vol. 1, 365–368.

10 *The Balanced Scorecard in Food and Farming*

NICOLA SHADBOLT
Massey University

Strategic Management

Strategic management differs from other levels of management in several ways: it is non-routine, non-programmable, unique and creative (Harrison, 1999), more ambiguous, uncertain and complex (Johnson et al., 2005) and yet it has the greatest impact on the future of the business (Shadbolt and Bywater, 2005). Strategy defines the logical case for how value will be created for shareholders; it will define actions and resource use but, inevitably, it is based on a set of assumptions about the future that must be put to the test.

Porth (2003) suggests that the strategic management process includes five interrelated tasks: to develop a mission and a vision, to perform a situation analysis (internal and external audit), to set objectives and to craft the strategy, to implement the strategy, and to assess value creation and provide feedback. There is significant debate in the literature as to what constitutes a mission and a vision; for family businesses this author prefers to take the Johnson et al. (2005) approach of ensuring the core values of the business (and family) are understood and reconciled before defining what they term the 'organisational purpose'. Family businesses frequently have stakeholders with opposing value sets, so it is important that those differences are recognised before a value set relevant to the business is devised.

Studies of farmers' goals and objectives discussed by Gasson and Errington (1993) identify that automony, independence, survival and succession mingle with more orthodox economic issues.

Examples of vision statements from New Zealand farms are given in Table 10.1. Note how clearly the differing value sets become obvious from the words they have chosen to use.

In a larger organisation it is expected that leadership is delivered by the CEO and the board of directors; management is delegated to the senior and middle managers and operational issues are delegated to staff at the 'coalface'. In smaller companies such distinctions, although still relevant, can be lost, and it is easy to understand how the most frequent issues that, by definition, will be operational ones can dominate over less frequent ones. There is a stronger focus at the operational and, sometimes, the tactical level of the business and a weak strategic focus (Shadbolt and Rawlings, 2001; Doonan,

Table 10.1 Examples of vision statements from New Zealand family farm businesses

To pursue excellence and growth in sustainable agriculture thereby providing for diversification, succession, and a continuously improving standard of living for all stakeholders in the business.
To have a farm business that is growing, is financially profitable and environmentally sustainable while enabling us to maintain our lifestyle and enjoy time with our children and friends and continue our involvement in the community .
To increase sustainable net income by being a preferred supplier of high-quality milk and dairy livestock, by adopting excellence in farm management practices and technologies and by developing beneficial partnerships with our team of staff, while embracing the values of integrity, honesty and the pursuit of knowledge.
To grow the family farm business by profitably marketing quality products to ensure that the future generation has a viable farming business.

2001). These businesses run the risk of falling into the trap of being very efficient at getting the wrong job done.

To belittle farm managers for having this short-term focus is to misunderstand the complexity of their role. Operational and tactical management are essential skills on all farms and must be mastered. If the business is small it is they who must master them, as delegation cannot occur. However, they must also deliver to the dream that they and their family have, so leadership skills are essential; recognising how their operational and tactical activities deliver to that dream by determining the cause–and-effect relationships that exist is an acquired skill.

The Balanced Scorecard (BSC) as a Key Part of Strategic Management

The shift from the industrial economy towards an economy characterised by intangible assets, such as knowledge and innovative capability, has increased the levels of complexity, mobility and uncertainty organisations face (Voelpel et al., 2006). Atkinson (2006) summarises that the transformation from the industrial age to the information age is signalled by increasingly sophisticated customers and management practices, escalating globalisation, more prevalent and subtle product differentiation and an emphasis on intellectual capital and enhanced employee empowerment. A range of new strategy tools and performance measurement frameworks have evolved to assist strategy implementation. Non-financial measures have combined with or replaced traditional finance-orientated metrics as strategic controls providing useful short-term targets on the long-term strategic road (Bungay and Goold, 1991).

One such tool, the balanced scorecard, is described by Atkinson (2006) as arguably the dominant framework in performance management. Devised by Kaplan and Norton (1992), it was proposed as an approach to tracking a firm's performance that takes into account process, innovation and customer objectives as well as the financial position. In working with the scorecard they also found it performed an integrative function by bringing together disparate measures in a single report, and hence helped the senior management team to clarify and operationalise strategy (Knott, 2006). They identified significant weaknesses in performance management systems at that time that were dominated by short-term, backward looking or 'lag' financial metrics, which were internally orientated and not linked to organisational strategy (Atkinson, 2006). Based on case study research of leading companies they concluded that financial numbers alone were no longer sufficient to run a business effectively because they lacked predictive power. They devised a scorecard with four perspectives that permitted a balance to be struck between short- and long-term objectives, between desired outcomes and the performance drivers of those outcomes, and between hard, objective measures and the softer, more subjective, measures (Haapsalo et al. 2006). In response to the tension that exists between the rigour necessary for effective strategy implementation and the flexibility required for timely strategic adjustment (Atkinson, 2006), they also claimed that the balanced scorecard 'provides a framework for managing the implementation of strategy while also allowing the strategy itself to evolve in response to changes in the company's competitive market and technological environment ' (Kaplan and Norton, 1996).

The BSC Framework

According to Kaplan and Norton (1996), a balanced scorecard should have the following components:

- *Core vision*: The value-based organisational purpose that strategy delivers to.
- *Perspectives*: There are typically four perspectives: financial, customer, internal processes, and learning and growth. Others may be added based on specific needs. A perspective often represents a stakeholder category or point of view.
- *Objectives*: An objective states how a strategy will be made operational. They usually form the building blocks for the overall strategy of the organisation.
- *Measures*: It must be quantifiable. They communicate the specific behaviour to achieve the objective and become the actionable statement of how the strategic objective will be accomplished. Lead measures are predictors of future performance (drivers), while lag measures are outcomes.
- *Strategic initiatives*: These activities (discretionary investments or projects) will focus on the attainment of strategic results. All initiatives in an organisation should be aligned with the strategy in the balanced scorecard.
- *Cause-and-effect linkages*: Similar to 'if–then' statements, these cause-and-effect linkages should be explicit.

Some authors suggest that having only four perspectives is a weakness in the BSC. Working with on-farm agribusinesses, Shadbolt et al. (2003) found an extension of the customer perspective to include the suppliers (more of a supply chain approach) was relevant, as was an extension of the shareholder/financial perspective to include non-financial shareholder goals such as lifestyle and environmental/ethical issues. Dunn et al. (2006) captured these issues by suggesting six perspectives suitable for ranch strategic management to ensure the lifestyle and environmental (natural resources) aspects of the business were given equal weight to the more traditional perspectives of financial, livestock production, customer, and learning and growth.

So the BSC allows executives to manage a company from several perspectives simultaneously. Shadbolt (2004) states it has evolved into a useful framework as it forces the perspectives of human resources (innovation, continuous improvement and learning), internal processes (turning inputs into outputs), the market (customer relationships, product and service criteria) and shareholders (profitability, return on assets, wealth, non-financial and ethical goals) to be explored and the linkages between them to be determined. The term 'balanced scorecard' reflects the balance between short- and long-term objectives, financial and non-financial measures, lag and lead indicators, and external and internal performance perspectives (Hepworth, 1998). It provides a balanced organisational assessment by recognising a variety of stakeholder views.

The BSC design process builds upon the premise of strategy as hypotheses. Strategy implies the movement of an organisation from its present position to a desirable but uncertain future position. Because the organisation has never been to this future position, its intended pathway involves a series of linked hypotheses. The scorecard enables the strategic hypotheses to be described as a set of cause-and-effect relationships that are explicit and testable (Kaplan and Norton, 2000). Shadbolt (2004) gives an example of this for on-farm agribusiness, describing how the assumptions made of the cause-and-effect relationship between process (farm practices) and state (environmental impacts) indicators could be explored.

Having a sound vision for the business is the key to the success of the BSC (Haapasalo et al., 2006). As already stated, a common vision is a challenge in farm family businesses, where conflict often exists between business and family visions and purpose (Gasson, 1973; Byles et al., 2002). A solution proposed by Andersson (2002) was separate visions for business and for farm family lifestyle issues and the addition of a fifth perspective to the BSC, called 'life'. However, to have two visions could be divisive and lead to family business dysfunction: Atkinson (2006) identifies from the literature that in all businesses, regardless of size, strategic change requires a shared vision and consensus; failures are inevitable if competence, co-ordination and commitment are lacking.

The absence of goals or abundance of goals in any perspective gives a quick, visual indication of whether the business is 'in balance'. Key metrics are also specified for each goal, and include both the outcomes (lag indicators) and the drivers (lead indicators). If too many metrics are defined in a BSC it quickly turns from a management system into a monitoring system; if it is to be used effectively as a management tool with strategic purposes the number of metrics must be low (Haapasalo et al., 2006).

Non-financial measures are usually drivers, informing the manager of likely future performance. For example, learning new knowledge and skills, a lag indicator for learning and growth, is a lead indicator of the farm staff's ability to ensure best practices at 'harvest' are in place (Shadbolt et al., 2003). Without investment in staff learning and personal growth, the business has less ability to deliver to the product quality specifications identified in its customer related goals. The under-utilisation of non-financial key performance indicators in business control was one of the key findings that led to the development of the balanced scorecard by Kaplan and Norton (1992).

Making the BSC Work

Atkinson's (2006) review on the role of the BSC in strategy implementation identifies some critical aspects to making it work:

1. Begin with a full strategic appraisal and a clear articulation of the organisation's strategic vision and objectives – this builds consensus and engenders learning.
2. Make explicit the cause and effect of a strategy – convert strategic aims into tangible objectives and measures and identify where they interlink.
3. Implement the BSC participatively with measures identified and targets set co-operatively rather than imposed – this supports organisational learning and reflection and encourages interactive control through the testing of 'cause-and-effect' relationships.
4. Encourage BSCs at every level of an organisation to enable middle management engagement.

Haapasalo et al. (2006) state there should be profound discussion over different perspectives. Again the advice is to determine what is right for each company – do not use other company or industry lists. When applied to on-farm agribusinesses (Shadbolt and Rawlings, 2001; Shadbolt et al., 2003) it has proven to provide an acceptable framework with which to capture the more holistic nature of farm systems and enable both financial and non-financial (including non-business) goals to be managed. Shadbolt (2004) further suggests how in a policy context it can provide a framework to enable a specific plan, such as an environmental plan, to fit within the overall business.

Figure 10.1 provides an example of a scorecard for a pastoral farm business in New Zealand. The simplicity of the scorecard should not be taken as being an indication of lack of in-depth discussion and debate on strategy and metrics, but instead is an illustration of how it can be a useful report that clarifies and operationalises strategy.

It illustrates the cause-and-effect relationships and how they build from each level. The starting point is learning and growth; without human capability and capacity strategy cannot be implemented, and this leads on to the management of natural resources and production through which market needs are met. The result of meeting these needs is that financial targets are met and this, along with other non-financial outcomes, delivers to the lifestyle requirements of the family.

Conclusion

It is important to recognise that a strategy tool like the balanced scorecard is likely to assist with part of the activity rather than provide a substitute for the capabilities and experience of the manager. It does not provide a blueprint but can act as a guide to thinking and a starting point for structuring the activity (Knott, 2006). There is a risk that the tool or framework that a manager uses will channel or constrain thinking as it focuses and guides, in which case alternative tools or adaptation of the tool may be required to ensure robust strategy is crafted. Users of the balanced scorecard should keep this in mind: the tool is not a recipe for success but a means by which a business can assess its direction, craft strategy and define success. Its application will vary from ranch to ranch according to the skills and motivation of the owners and managers: for some it might just guide thinking and debate; for others it will provide a framework that will enable greater detail on implementation and on-going control.

The framework the balanced scorecard provides enables the components of strategy to be identified and the interactions between the components to be visualised. It forces goals to be linked to the vision and actions to be linked to goals. It enables ranch businesses to look beyond finance and production to include multiple perspectives. Farmers must not only define how success is measured in the balanced scorecard but also what the drivers of positive change really are, what are the cause-and-effect relationships in the business. And, finally, to make best use of the balanced scorecard, keep it flexible, simple and practical.

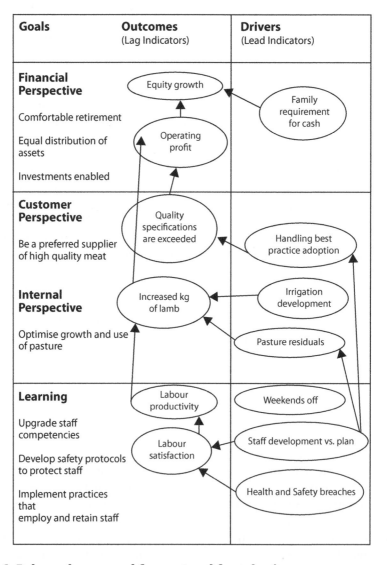

Figure 10.1 Balanced scorecard for pastoral farm business

Source: Martin and Shadbolt (2005).

References

Andersson, P. (2002), Competence development program for the farmer with reference to life as well as business, *Proceedings of the 13th International Farm Management Association Congress, The Netherlands July 7–12, 2002.*

Atkinson, H. (2006), Strategy implementation: a role for the balanced scorecard?, *Management Decision* Vol. 44 No. 10, 1441–1460.

Bungay, S. and Goold, M. (1991), Creating strategic control systems, *Long Range Planning* Vol. 24 No. 3, 32–9.

Byles, S., Le Grice, P., Rehman, T. and Dorward, P. (2002), Continuing professional development and farm business Performance. *Proceedings of the 13th International Farm Management Association Congress, The Netherlands July 7–12, 2002.*

Doonan, B.M. (2001), Strategic planning in the dairy industry – the Tasmanian experience. *Proceedings of the South Africa Large Herds Conference. Port Elizabeth, 2001.*

Dunn, B.H., Gates, R.N., Davis, J. and Arzeno, A. (2006), *Using the Balanced Scorecard for Ranch Planning and Management: Setting Strategy And Managing Performance* (South Dakota State University/Texas A&M University-Kingsville) (EC922) <http://agbiopubs.sdstate.edu/articles/EC922.pdf>.

Gasson, R. (1973), Goals and values of farmers, *Journal of Agricultural Economics* Vol. 24, 521–537.

Gasson, R. and Errington, A. (1993), *The Farm Family Business* (Wallingford: CAB International).

Haapasalo, H., Ingalsuo, K. and Lenkkeri, T. (2006), Linking strategy into operational management, *Benchmarking: An International Journal* Vol. 13 No. 6, 701–717.

Harrison, E.F. (1999), *The Managerial Decision Making Process.* (Boston, MA: Houghton Mifflin).

Hepworth, P. (1998), Weighing it up – a literature review for the balanced scorecard, *Journal of Management Development* Vol. 17 No. 8, 559–563.

Johnson, G., Scholes, K. and Whittington, R. (2005), *Exploring Corporate Strategy: Text and Cases.* 7th edition (London: Pearson Education Limited).

Kaplan, R. S., and Norton, D. P. (1992), Using the balanced scorecard as a strategic management system, *Harvard Business Review* Vol. 70 No. 1, 71–80.

Kaplan, R. S., and Norton, D. P. (1996), The balanced scorecard – measures that drive performance, *Harvard Business Review,* Vol. 74 No. 1.

Kaplan, R.S. and Norton, D.P. (2000), *The Strategy-Focused Organization: How Balanced Scorecard Companies Thrive in the New Business Environment* (Boston: Harvard Business School Press).

Knott, P. (2006), A typology of strategy tool applications, *Management Decision* Vol. 44 No. 8, 1090–1105.

Porth, Stephen J. (2003), *Strategic Management: A Cross-functional Approach* (New Jersey: Prentice Hall).

Shadbolt, N.M. (2004), Agri-environmental indicators put into perspective: their fit and relationship with other relevant farm business indicators, In Fraser, N. (ed.) *Farm Management Indicators and the Environment, Proceedings of an OECD Expert Meeting, Palmerston North, New Zealand, March 2004.*

Shadbolt, N.M., Beeby, N., Brier, B. and Gardner, J.W.G. (2003), A critique of the use of the balanced scorecard in multi-enterprise family farm businesses. *Proceedings of the 14th International Farm Management Congress: Pat 1, 602–609. August 10–15, 2003. Perth, Australia.*

Shadbolt, N.M. and Bywater, A. (2005), The Dimensions of Management. In Shadbolt, N.M. and Martin, S. (eds) *Farm Management in New Zealand* (Oxford: Oxford University Press).

Shadbolt, N.M. and Rawlings, M. (2001), *Successful Benchmarking by Balanced Planning and Identifying Key Performance Indicators for Goal Attainment in Dairy Farming.* Dairy Research and Development Corporation (Australia). Project Code: MUNZ001.

Voelpel, S.C., Leibold, M and Eckhoff, R.A. (2006), The tyranny of the Balanced Scorecard in the innovation economy, *Journal of Intellectual Capital* Vol. 7 No. 1, 43–60.

11 Conclusion: Creating Sustainable Change through Benchmarking in Food and Farming

Change is a process and as such should requiring planning, management and maintenance in order to be sustainable. The case studies and examples given throughout the book mainly concern producers, as that is where most evidence of benchmarking for best practice in the industry is to be found. The reasons for change are frequently related to establishing a firmer position in the food supply chain, or to finding routes to market outside the chain. Some changes are to meet conditions imposed by supply chain organizations downstream. Change processes taking place throughout the whole supply chain are rare. What can be seen is that successful, sustainable change comes through planning and monitoring change processes, and that benchmarking plays an important role.

Benchmarking in agriculture is well established, and in the case of mass participation benchmark systems, institutionalized. Benchmarking in other areas of the food chain is more ad hoc: there are instances of internal benchmarking, such as the cases of Frito-Lay and Kellogg's given in Chapter 9, but attempts by certain advisory companies to develop competitive benchmarking in the industry have had a low take-up. Techniques such as efficient customer response (ECR) and supply chain logistics include key performance indicators – particularly non-financial ones – but whether these are actively used for benchmarking is debatable. Benchmarking, as seen in Chapter 1, implies a significant element of learning, information sharing and adaptive innovation of best – or at least better – practices, not simply monitoring of performance.

Given that farm benchmarking practices are well established but those in the rest of the food chain are less established, the question is whether benchmarking has the potential to create sustainable change throughout the food chain. The question falls into four parts:

1. Should farmers be encouraged to adopt more competitive forms of benchmarking?
2. Should distributors, processors, wholesalers and retailers be encouraged to make more use of competitive benchmarking?
3. Can benchmarking be used in food supply chains?
4. In all these areas, what is the role of environmental benchmarking?

As highlighted in Chapter 1, there are issues surrounding the sharing of information and of trust between participants in food supply chains, not to mention significant complexity and fragmentation, and these inhibit benchmarking. Yet this business environment suggests innovations and creative solutions are needed, and that non-financial as well as financial-based benchmarking has potential to address some of these issues.

Competitive Benchmarking in Farming

In Part 1, mass participation benchmark systems for farming were seen to provide aggregate benchmarks that were useful for policy setters, advisors and researchers and so contribute to the information that created higher level change and widespread advice for farmers. Whilst individual farmers may get a report from the survey data collectors, or may be able to filter a like-for-like report from the database, there is very little empirical evidence to indicate that these figures motivate farmers to make significant changes. There are statistical databases for food businesses, largely run by governments, but again there is almost no evidence that any benchmarks generated by these systems are used in any significant way by food companies. However, there is evidence that newer mass participation benchmark systems are being developed for specific industries, with links to other participants in the food chain. Examples of this are the DairyBase system in New Zealand discussed in Chapter 4 and the schemes identified in Chapter 3 where there was data sharing with supply chain partners such as feed merchants or lenders. In the latter cases, it was reported that benchmarking added detail to farmers' discussions and negotiations with their supply chain partners in a way that was beneficial. Opportunities to package advisory services to farmers were being created through the provision of benchmark services.

The key to the future development of mass participation benchmarking is for data from a number of supply chain partners, not just farmers, to be included. As discussed in Chapter 3, if all partners are involved – and technologies now make this increasingly a possibility – and the information becomes a necessary part of doing business with each other, then other food chain members will be obliged to participate or will lose competitive advantage. There needs to be advocacy for these systems, and less sensitive data may be the first to be shared, but with support there is a potential to develop benchmarking systems that are of mutual benefit to all participants. This point is discussed further in the section relating to benchmarking in the food chain.

The other developed area of benchmarking in farming is the discussion group, which originated in New Zealand in the 1950s and has spread more recently to Australia, the UK and Ireland. There are other forms of advisory groups in the US and Europe, but the discussion group has become a more effective way of using benchmarking. It is effective because farmers do, as Bogan and English (1994) suggest, find not only stories but also physical evidence of successful change, and a supportive environment in which to make decisions. The format has also been adapted in Australia to work on environmental targets and is evolving elsewhere to look at separate key performance indicators such as machinery and labour costs, and veterinary problems.

There is no evidence that other food chain companies meet in this way, and in fact outside agriculture the network or learning benchmark group is seen as a twenty-first century development. Again, this point is returned to later in discussing food chain partners. For farmers, the most innovative benchmarking is moving towards what would be called elsewhere 'competitive benchmarking' and in Chapter 9 examples are given of internal benchmarking and near-process benchmarking. Extending this further, to seek best practice on a one-to-one basis and overseas is likely to be the logical and desirable development of benchmarking. Therefore, it can be said that farming is already moving towards using competitive benchmarking and that advisers, consultants and policy makers should be facilitating these practices.

Competitive Benchmarking in Food Companies

There are examples of internal benchmarking in food companies but very little publicly available information concerning other benchmarking practices. However, research suggests that as management accounting practices generally in the food chain are less well developed than might be expected, it is reasonable to assume that competitive benchmarking is not an established feature of management practice in the industry. It is difficult to know without carrying out new research whether this is because it has been tried and found to be unsuccessful, or whether it has not been championed in the same way as other management tools. What is plausible is that the increasing concentration of food companies – only a small number of retailers, processors or distributors accounting for over 70 per cent of activity in some food chains – has firstly decreased the number of potential partners and secondly increased the sensitivity of information held by each. Hence, competitive benchmarking becomes more of a risky proposition to strategic managers.

Process benchmarking in other industries, however, has looked outside a company's own sector for best practice and there are indications that this is happening in the field of supply chain logistics in the food chain. Internal benchmarking does happen, as seen in Chapter 9 and there are examples of the balanced scorecard (BSC) being used by some food companies. Tesco, the UK retailer, claims to have been using BSC since its introduction by Robert Kaplan in the early 1990s. The company uses an adaptation called the Steering Wheel, with community as an added fifth area (www.tesco.com/). The extent to which the tool is used for benchmarking is less publicised but in an article in *Personnel Today* magazine (Paton, 2005) it is stated that:

Tesco uses a balanced scorecard approach to management through its 'Steering Wheel' programme. Managers monitor customers, operations, staff and finances using a traffic light system to denote meeting targets and finding problems.

This is an internal strategic management tool, and it is less evident to what extent it is used to manage food chain partners. Other examples of food companies using BSC for internal improvements include restaurant chain Wendy's and other fast-food companies, with an emphasis on quality and customer service. Interestingly, however, a report on the use of BSC by Wendy's does not mention suppliers or raw ingredients (Ross, 2003). The use of the BSC to translate strategy into targets is also used by Summa Health Food Systems (Smith and Koon, 2005). The challenge is to extend the BSC across the supply chain as a 'multi-enterprise' BSC. The area in which food chain companies can most usefully extend the concept of competitive benchmarking appears to be in managing the supply chain.

Competitive Benchmarking in the Food Supply Chain

Researching food supply chains uncovers some surprising facts for the outsider. There are very few written contracts between producers and their customers downstream. Information sharing is unusual, at least on a voluntary basis, and what is shared is non-financial. This may seem to be prudent on behalf of the participants but it also inhibits the possibility of ascertaining whether participants are being awarded a fair share of the value created along the supply chain. Statistics indicate that whilst the value added to food has increased significantly in the last decade, the return to primary producers has stayed the same (and in fact decreased in real terms).

One article highlights five steps to getting a multi-enterprise BSC, which could apply to any benchmarking scheme throughout the food supply chain (Kent and Geraghty, 2004): a common language, rationalized measures, supply chain targets, data management and transparency. The SCOR (Supply Chain Operations Reference) provides a common language and systems which could manage the data, but what is lacking, it seems, in food supply chains is firstly the commitment to create measures and targets from within the industry that are acceptable by all parties (there are a number of academic solutions that have yet to gain currency) and more importantly, a culture of transparency.

Benchmarking in this context could develop on a number of fronts, as it has among producers. Mass participation systems could provide guidance and 'reality checks'; discussion groups between supply chain partners could facilitate exchange of knowledge and foster trust and transparency; competitive, process benchmarking could be used to identify opportunities in the distribution and processing of food, where it is acknowledged that there is waste and undue 'food miles' being clocked up, and data could be gathered to indicate where value is being delivered and where not; and internal benchmarking could be extended to include suppliers.

Competitive Benchmarking for the Environment

The challenge in all the benchmarking activities that have been discussed is how social and environmental factors should be included. Are they one set of indicators, an added sector of a scorecard or separate benchmarking system, or can they be integrated? The term 'sustainable change' has been used throughout this book in the sense of making changes that will enable businesses to last and where the changes themselves can be sustained. However, the implication is that unless these changes are not damaging to environment and society, they cannot be sustainable. In Chapter 7 the range and complexity of environmental benchmarks in the form of targets, codes and certification schemes was explored, and whilst farmers and food companies do make changes to meet environmental demands, these are largely customer driven.

A problem seems to be the setting of achievable targets in this area, given the large quantity of information available, combined with the difficulty of having clear outcomes to benchmark against. By definition, sustainability refers to future events we cannot fully model. Yet the improvement groups pioneered in Australia which are reported in Chapter 8 seem to hold a key: if benchmarking is about learning and innovation, then perhaps practical solutions can be found through discussion, evaluation and monitoring. This could apply also to food supply chains, although evidence again shows that coercion or incentives have been hitherto the most effective drivers of change.

There are stories from producers and others of changes made because of benchmarking activities that have made businesses stronger, more viable and more profitable, and where farmers have admitted to having increased their understanding of costs and activities from going through the processes of benchmarking. There is scope for extending this learning throughout the food supply chain, and benchmarking can be one tool for creating sustainable change in the industry.

References

Bogan, C.E. and English, M.J. (1994), *Benchmarking for Best Practices: Winning Through Innovative Adaption* (New York: McGraw-Hill).

Kent, D. and Geraghty, K. (2004), *Extend your Balanced Scorecard Across the Supply Chain* <http://www.performance-measurement.net/news-detail.asp?nID=199>.

Paton, N. (2005), Supermarket sweep: Tesco, *Personnel Today* 5 July 2005.

Ross, J.A. (2003), The best practice hamburger: how Wendy's enhances performance with its BSC, Balanced Scorecard Report, Article reprint No. B0307B (Boston: Harvard Business School Publishing).

Smith, H. and Koon, I. (2005), Balanced Scorecard at Summa Health System, *Journal of Corporate Accounting & Finance* Vol. 16 No. 5, 65–72.

Index

Printed and bound by CPI Group (UK) Ltd, Croydon, CR0 4YY

18/10/2024

01776204-0012